THE
PORTABLE
ENGINE

Its Construction and Management

A PRACTICAL MANUAL FOR OWNERS AND USERS OF STEAM ENGINES GENERALLY

BY

W. D. WANSBROUGH

Author of " The A B C of the Differential Calculus," " The Proportions and Movement of Slide Valves,"
&c. &c. &c.

With 118 Illustrations

TEE Publishing
Warwickshire, England

First published 1911
Reprinted 1994
© TEE Publishing ISBN 1 85761 067 9

PREFACE

THE author acknowledges with gratitude the reception afforded to his earlier volume, "The Portable Engine: Its Construction and Management." In response to the demand for further and more modern treatment of the subject he has thought it best to re-write the work entirely, and to issue a new book under a new title.

Much of the matter contained in the original work and its reprints had become more or less obsolete in the natural order of things, and some of the more elementary portions, it was felt, might well be omitted in favour of more useful and practical information. The author is glad therefore of the opportunity now afforded of bringing the matter quite up to date through the kindly assistance of the various leading makers of the portable steam engine, whose productions are here set forth. He only regrets that it has been impossible to include within the assigned limits any notice of the kindred types of semi-portable and under-type engines, the enormous development of which during the last few years would, if adequately dealt with, demand another volume of size at least equal to the present one.

In the Historical Section the author has attempted to trace the Evolution of the Portable Steam Engine. The result of his researches may, or may not, "fill a long felt want"; but at any rate he is enabled to place on record, for the first time in anything like a permanent form, the slow and hesitating steps, from the days of Trevithick onwards, by which the movable steam engine has reached its present position and place in the service of mankind.

Finally, the author here records his sincere thanks to the editors of *Cassier's Magazine* and of *Page's Magazine* for their readily accorded kind permission, enabling him to make use of papers on the subject contributed by him from time to time to their respective magazines—in the latter case under the pseudonym of "J. C. R. Adams," and anonymously.

Exchange Buildings,
Birmingham, *December* 1911.

CONTENTS

CHAPTER I
THE EVOLUTION OF THE PORTABLE STEAM ENGINE - - - - 1   PAGE

CHAPTER II
THE PORTABLE STEAM ENGINE OF TO-DAY - - - - - 27

CHAPTER III
SOME STANDARD TYPES OF SINGLE-CYLINDER ENGINES - - - 39

CHAPTER IV
THE COMPOUND PORTABLE ENGINE - - - - - - 65

CHAPTER V
SOME SPECIAL TYPES OF PORTABLE ENGINE - - - - - 96

CHAPTER VI
PRACTICAL HINTS ON USE AND MANAGEMENT - - - - 109

CHAPTER VII
THE SLIDE-VALVE AND ITS ACTION - - - - - - 130

CHAPTER VIII
THE INDICATOR DIAGRAM - - - - - - - 150

INDEX - - - - - - - - - 167

LIST OF ILLUSTRATIONS

FIG.		PAGE
1. Trevithick's High-pressure Engine, 1811		4
2. Tuxford's Combined Engine and Thrasher, 1839-1842		6
3. Ransomes' ,, ,, 1841		7
4. Dean's Portable Engine, 1844		9
5. Cambridge's Portable Engine, 1847		10
6. ,, Boiler (two views), 1847		11
7. Hornsby's Portable Engine, 1848		13
8. Bach & Co.'s ,, 1850		14
9. Tuxford's ,, 1850		15
10. ,, ,, (end view)		16
11. Hornsby's ,, 1851		17
12. Garrett's ,, 1851		·18
13. Ransomes' ,, 1851		19
14. Clayton's ,, 1853		20
15. Robey's ,, 1861		20
16, 17. Robey's Crankshaft Governor, 1869		21
18-21. Indicator Diagrams		24-26
22. Firebox Roof Girders, Old Style		32
23. Garrett's Corrugated Roof Plate		34
24. Robey's Roof Girders		35
25. Marshall's Firebox Crown		36
26. Brown & May's Portable Engine		40
27, 28. Clayton & Shuttleworth's Portable Engine		42, 43
29. ,, ,, Cylinder		44
30. ,, ,, Plummer-block		45
31. ,, ,, Eccentric		46
32. ,, ,, Cylinder and Governor		47
33. Davey, Paxman, & Co.'s Portable Engine		48
34. ,, ,, Governor		49
35. Garrett's Portable Engine		To face 54
36. ,, Flanged Steel Saddle		52
37. ,, Plummer-block		53
38. ,, Cylinder		53
39. ,, Governor		54
40. ,, Feed Pump		54
41. Marshall's Portable Engine		To face 55
42. ,, Governor		56

LIST OF ILLUSTRATIONS

FIG.		PAGE
43. Ransomes' Portable Engine		58
44. Robey's Portable Engine		59
45. ,, ,, (old pattern)		To face 60
46. Ruston, Proctor, & Co.'s Portable Engine		,, 61
47. ,, ,, Cylinder		61
48. ,, ,, Eccentric		62
49. Turner's Portable Engine		63
50. Brown & May's Compound Portable Engine		69
51. Clayton & Shuttleworth's Compound Portable Engine		70
52. Davey, Paxman, & Co.'s ,, ,,		72
53. ,, ,, Diagrams		73
54. ,, ,, Expansion Gear		74
55. Wm. Foster & Co.'s Compound Portable Engine		76
56. ,, ,, Diagrams		77
57. Garrett's Compound Portable Engine		To face 78
58. ,, ,, ,,		,, 79
59. ,, Cast-iron Saddle		79
60. ,, Expansion Gear		80
61. Marshall's Compound Portable Engine		To face 80
62. ,, Governor		81
63. Ransomes' Compound Portable Engine		83
64. ,, Governor		84
65. Robey's Compound Portable Engine		86
66. ,, ,, ,,		87
67, 68. Robey's Automatic Expansion Gear		88, 89
69. Robey's Diagrams		90
70. ,, 200 H.P. Compound Portable Engine		To face 90
71. Ruston, Proctor, & Co.'s Compound Portable Engine		91
72. ,, ,, Rider Expansion Gear		92
73. ,, ,, Diagrams		94
74. Robey's Compound Portable Engine with Condenser		97
75. Ruston, Proctor, & Co.'s ,, ,,		98
76. Garrett's Jet Condenser for Portable Engines		99
77. ,, Superheated Steam Single-cylinder Portable Engine		100
78. ,, Special High-pressure Portable Engine		To face 91
79. Marshall's High-pressure Portable Engine		,, 102
80. Ruston, Proctor & Co.'s High-pressure Portable Engine		,, 102
81. Brown & May's Portable Engine with Circular Firebox		,, 102
82. Marshall's ,, ,, ,,		,, 102
83. Robey's Portable Engine with Circular Firebox		104
84. Ruston, Proctor, & Co.'s Portable Engine with Circular Firebox		105
85. Robey's Removable Firebox Boiler		106
86. Biddell & Balk's Patent Removable Firebox Boiler, 1858		107
87, 88. Diagrams of Crankpin Pressures		113
89. Automatic Expansion Gear		118
90. Effect of Connecting Rod		119
91. Indicator Diagrams		120
92. Automatic Governor and Expansion Gear		121
93. Section of Cylinder and Steam Chest		130

LIST OF ILLUSTRATIONS

FIG.		PAGE
94.	Diagram of Crank and Eccentric	131
95.	,, ,, (with lap)	132
96.	,, ,, (with lap and lead)	133
97.	Section of Slide-Valve and Cylinder Face	134
98.	Zeuner Diagram	138
99.	Slide Valve with Exhaust Lap	140
100.	Zeuner Diagram for Exhaust Lap	141
101.	Effect of Connecting Rod	142
102.	Zeuner Diagram Corrected for Error of Connecting Rod	143
103.	Slide-Valve Section	144
104.	,, ,,	144
105.	,, ,,	145
106.	,, ,,	145
107.	,, ,,	146
108.	,, ,,	146
109.	,, ,,	147
110.	,, ,,	147
111.	Zeuner Diagram for Variable Expansion	148
112.	Real and Imaginary Indicator Diagrams	151
113.	Diagram from a Locomotive Engine	154
114.	,, showing Method of Measuring	155
115.	,, ,, Unequal Valve-setting	156
116.	Seven Stages in the Tracing of an Indicator Diagram	159
117.	Sectional View of Indicator	160
	Tail-piece "The End"	165

The publishers and printers regret that the reproduction of some illustrations is less than they would like to have achieved but this is due to the poor quality of the illustrations contained in the original book of 1911.

THE PORTABLE STEAM ENGINE

CHAPTER I

THE EVOLUTION OF THE PORTABLE STEAM ENGINE

THE first step in the direction of making engines portable was undoubtedly the plan of using steam of such comparatively high pressure as to obviate the necessity of forming a vacuum, and the cumbrous apparatus of the condenser. At a very early date—1727—this appears to have occurred to Leopold of Plunitz, Saxony, who published in that year a Latin treatise in which he describes a double-cylinder engine wherein the admission and exit of the steam were regulated by a four-way cock, which diverted the current of steam to each cylinder in turn, at the same time opening communication with the atmosphere at the completion of each stroke. The piston rods appear to have been attached to beams for pumping. Of course there was no rotary motion, and it is extremely doubtful whether this engine was ever really constructed.

In the year 1765, Smeaton, in his "Reports," describes a movable engine and boiler, but as this appears to have been a condensing engine, with a cylinder of 6 ft. stroke, it is hardly necessary to go into detail concerning it.

The boiler, however, seems to have been self-contained and internally fired, and therefore has some claim to be considered portable. When the applicability of the steam engine to other purposes besides pumping became evident, many different methods were proposed for changing the reciprocating motion of the beam into rotative motion, some of them of extraordinary complexity. In justice to these early inventors, we must remember that the primary object of the steam engine at this period of its history was the pumping of water from mines; hence the ponderous beam which remained a feature alike of rotative and non-rotative engines for many years to come.

In 1779, Matthew Wasbrough, of Bristol (an ancestor of the present writer), patented three ingenious and complicated devices for converting reciprocating into circular motion, and shortly afterwards the crank, as we now know it, was applied to one of Wasbrough's engines at Birmingham by the purchaser, instead of the original ratchet motion. For this a patent was granted to John Pickard in 1780. Watt, disappointed by his failure to establish a prior claim to the crank, put forward in its place, in 1781, the sun-and-planet motion, which consisted of a spur-wheel fixed on the shaft, and another fixed to the connecting rod, a loose crank, or radius bar, of length equal to the pitch circle diameter, keeping the two in gear. It will be seen that with this arrangement the shaft will make two revolutions for each reciprocation of the engine. Considering that the common crank was an appurtenance of every turning-lathe at that time in use, it is to us marvellous that Watt allowed himself to be forestalled in its application to steam power.

The next important improvement which claims our attention as leading up to the possibility of the portable engine, is the invention by Murdoch, Watt's pupil, in 1785, of the slide-valve; and again, the first mention of a metallic piston packing appears to be that made in the application of Edward Cartwright for a patent, in 1797, previous pistons (and later ones too) having been packed with hemp.

In a patent granted to Matthew Murray in 1802, what appears to be really a portable engine is described, and even its suitability for agricultural purposes indicated. After a description of the details of the engine, the specification proceeds as follows: " The parts so combined form a perfect engine, without requiring any fixture of wood, or other kind of framing than the ground it stands upon; and is transferable without being taken to pieces. The motion of the flywheel gives circular power to any process or manufacture requiring circular motion, or for irrigating land, and for the various purposes of agriculture."

Here we must break off for a moment to call attention to the singularly clear and explicit wording of Murray's claim. It was for an engine which is "transferable without being taken to pieces." At a period when steam engines, even of small power, were commonly incorporated into the structure of the house, which formed at once the framework and shelter of the machine, and were consequently about as portable as a parish church, this invention of Matthew Murray's was a long step in our direction. But an even greater one was soon to be made by Richard Trevithick, who has been well called the apostle of high-pressure steam. Trevithick was born in 1771, his father being himself a mining engineer of considerable eminence, who did much towards the improvement of the old atmospheric engine, and proved a rather formidable rival to Watt himself in the county of Cornwall.

We cannot resist dwelling for a few moments upon the inventions of this extraordinary and erratic genius. Listen to his own description of one of his engines:—"Chapell's engine had two cylinders and a double crank: the engines were fixed on the boiler; the piston rod crosshead worked in guides fixed to the cylinder; connecting rods went from the crossheads to the cranks. . . . Steam was turned on and off (*i.e.*, was admitted to each end of the cylinders alternately) by a four-way cock." This engine, it seems, was furnished with a blast-pipe, but whether this was used to cause an artificial draught, does not appear. At any rate it is described as "puffing so loudly that it could be heard for miles." In the account given of another of Trevithick's engines about the same period, however, it is explicitly stated that "the steam puffed up the chimney"—the term "blast-pipe" being then unknown to describe the thing just brought into use. But the words "the steam continued to rise the whole of the time it worked until it was necessary to stop and put a cock into the mouth of the discharging pipe to allow some of the blast steam to escape into the feed-warmer," prove the invention of the blast-pipe with its regulating cock and transmission of heat from surplus steam into the feed water to have been fully understood by Trevithick.

It seems difficult to believe that the description just quoted was written in the year 1802 at a time when Cornish engines on Watt's system were still working with boilers wholly or partially of *stone*, and engines dependent in most cases upon the masonry and walls of the engine-house for the connection between the beam and the cylinder. Trevithick's engines were, moreover, direct-acting (*i.e.*, having no beam) so that we have at this early date the essentials of the portable engine—high pressure, direct action, and forced (or rather induced) draught. The boiler appears to have been of cast iron, cylindrical, with a single wrought-iron tube passing through it, quite independent of brick-work foundations or setting, and the engine was attached to the boiler. The steam pressure would seem to have been about 30 lbs. per square inch.

These high-pressure engines were vehemently opposed by the firm of Boulton & Watt, at that time all-powerful in the new industry of steam engine making, who tried to get an Act of Parliament to prevent the construction of any more of them on the plea that the lives of the public were endangered. One of Trevithick's engines, built in 1811 for Sir Christopher Hawkins, was a prominent exhibit at the Royal Agricultural Show at Kilburn in 1879 (Fig. 1). It was taken from work where it had been intermittently in use for sixty-eight years, and was to be again put into steam on its return.

It is to be feared that no modern portable engine is likely to enjoy such an extended term of life as this venerable machine has done. In the "Life of Trevithick" there are many other instances of the extraordinary length of time that these engines have been

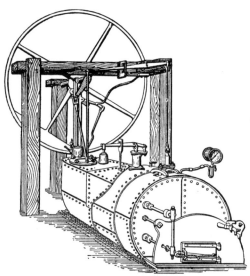

Fig. 1.—Trevithick's High-pressure Engine, 1811.

kept at work. In 1812 Trevithick was in a position to advertise publicly that he was prepared to make and supply portable engines for agricultural purposes, mounted on wheels and weighing 15 cwt., at a price of sixty guineas. It is necessary to resist the temptation of lingering among the memoirs of Richard Trevithick, one of the most original and daring inventors who ever drew breath. His high-pressure engine, condemned by Watt as dangerous and wasteful, gave to the small consumer, for the first time, the possibility of having steam-power upon his own premises at something like a moderate cost.

Trevithick's labours in connection with the locomotive engine, the screw propeller, steam ploughing and digging, and steam traction on common roads, do not concern us just now, but as an encouragement to future inventors a short extract from the closing pages of his "Life," written by his son, Mr Francis Trevithick, may be of service. Trevithick died at Dartford, in Kent, on 22nd April 1833. "He was penniless, without a relative by him in his last illness, and for the last offices of kindness was indebted to some who had been losers by his schemes. The mechanics from the works of Messrs Hall were the bearers and mourners at his funeral, and at their expense night watchers remained by the grave to prevent body-snatching, then frequent in the neighbourhood. His grave was among those of the poor buried by charity and no stone or mark distinguishes it from its neighbours."

During the next twenty or thirty years comparatively little progress was made with the portable engine. The impetus given by Trevithick to its manufacture did not, for some reason or other, continue long. It is probable that an explanation of this may be found in the disturbed state of the country during this period, and the high price of corn, which, combined with the growing dislike to machinery, as evidenced by the continual rioting and machine breaking in the manufacturing districts, would all operate to discourage the use of steam engines on the farm. A farmer in those days, employing machinery to any extent to do the work of men's hands, would be very likely to be reminded, by blazing ricks and maimed cattle, that there were thousands of starving labourers whom he would do well to employ, instead of adding to the distress by throwing more men out of work.

The business of making engines after Trevithick's designs was chiefly scattered among small foundries, such as would be established for the manufacture of ploughs and the smaller class of agricultural implements. No one firm took the lead in the matter, and thus, after a few engines had been made here and there at irregular intervals, the industry appears to have died out altogether.

Fixed condensing engines seem after this to have been used to some extent for agricultural purposes, chiefly in Scotland (where fixed thrashing machines are still employed to a much larger extent than in England), but we hear hardly anything further of anything approaching the portable engine until 1839, the date of the establishment of the Royal Agricultural Society. In that year Messrs Tuxford, of Boston, designed the curious machine shown in Fig. 2. This was finished in 1842, and set to work with satisfactory results.

Fig. 2.—Tuxford's Combined Portable Engine and Thrashing Machine, 1839-42.

It will be seen that the flywheel shaft runs at right angles to the crankshaft, driven by bevil gearing, and in its turn drives the drum-shaft of the thrasher by spur gearing. The cylinder is oscillating, and works the valve by its own motion. The fire-door is at the rear end, and the oval flue of the boiler returns to the chimney at the same end. A copper water-heater using exhaust steam forms part of the equipment. The engine was rated at 6 H.P., and the total weight, including the thrasher, was $3\frac{1}{4}$ tons. Within a twelvemonth from the trial of this engine, Messrs Tuxford turned out six more of the same class, and afterwards about twelve others of similar construction, after which they found it better to separate the engine from the thrashing machine.

In 1841 the Royal Agricultural Society held their meeting at

Liverpool, where an engine upon wheels was exhibited for the first time, and with this the history of the modern portable steam engine may be said to commence. This was a rotary disc engine, made by Messrs Ransomes & Sims, of Ipswich, under the patent of Mr Henry Davies, of Birmingham. As will be seen from Fig. 3, the engine was mounted upon four wheels and was self-propelling, the power being transmitted to the hind axle by means of a pitch chain. Upon the same platform was carried the thrasher, which was dismounted for work, and driven by a belt from the flywheel of the engine.

In the report of the judges it is stated to the credit of this machine that "it has no beam, flywheel, parallel motion, guide rods, condenser, air pump, or other intricate mechanism." With the exception of the flywheel, this is all undeniably true, as a single glance will show; and, encouraged by this eulogy, a company was formed at Birmingham for introducing it on a large scale; but, as is sometimes the case even now, the company got into difficulties, and we hear no more of the disc engine.

Fig. 3.—Ransomes' Combined Engine and Thrasher (Self-propelling), 1841.

The work done by this engine was estimated at 5 H.P., the quantity of water evaporated per hour at 36 galls., and the consumption of coke as 56 lbs. According to these figures the steam used per horse-power was 72 lbs. per hour, and the evaporative efficiency of the boiler was 6·4 lbs. water per lb. of coke, a result which shows that the boiler was of much greater efficiency than the engine.

It is perhaps needless to remark that at this date, and for many years to come, the steam engine trials of the Royal Agricultural Society were not by any means the models of painstaking accuracy that they have become in recent years, but enough was elicited from the results at this trial to show a very considerable saving to the farmer over the system of thrashing by hand.

The fact of the very high commendation of the engine by the Society did not, however, in the year 1841, remove the prejudice entertained by farmers against the use of steam-driven machinery in rick-yards; and the consequence was that Messrs Ransomes' machine had to be brought back from Liverpool, and was ultimately separated; the thrasher going to one purchaser, and the engine to another.

For some years after this the exhibition of portable engines at the Society's shows seems to have been mainly in the hands of two makers, whose names, we believe, have long since disappeared from the list of manufacturing engineers. These were Mr Alexander Dean, of Birmingham, and Mr W. Cambridge, of Market Lavington, near Devizes. Mr Dean is said to have made what may be termed a steeple engine in 1841. This he exhibited at Bristol in the following year, but no details of construction are now available. He had for competitors the Mr Cambridge just mentioned, with a portable engine having an oscillating cylinder, and Messrs Ransomes, of Ipswich, with a modification of their combined engine and thrasher. In this year also Messrs Clayton & Shuttleworth, of Lincoln (a firm which has since attained the highest eminence in this industry), began to make portable engines, their output being two for the year.

At the Society's meeting at Derby in 1843 three portable engines were shown by Dean and by Cambridge; and in reference to this the judges announced in their report, with great complacency, that "the manufacture and use of travelling steam engines has now become a systematic business." But it must not be assumed that these were anything like the trim and symmetrical machines now associated with the title of portable engine. Cambridge's was probably a vertical engine with the cylinder immersed in a cylindrical boiler with a return flue, but no authentic description exists of it. Dean's engine was vertical with a dome-topped boiler, and a rather ingenious arrangement of engine. The cylinder stood upright upon the bed-plate and the crank was carried overhead in a kind of square frame with pillars. Guides were cast upon the sides of the cylinder having sliding blocks connected by an arched crosshead with the piston rod. The connecting rod was of the forked type—very much so—being wide enough in the fork to embrace the arched crosshead just mentioned.

This made an exceedingly compact engine, the crank only just clearing the top of the crosshead.

In the year following—1844—the show was held at Southampton, with our friends Dean and Cambridge again as sole competitors as regards portable engines; but neither of these makers seems to have produced a very satisfactory engine. Dean's, which is the one shown in Fig. 4, was pronounced by the judges "not only inefficient, but highly dangerous"; while Cambridge's engine is stated, on the same authority, to have "consumed twice as much fuel as had been guaranteed by the maker." Evidently these judges were men of few words, but, nevertheless, they awarded a prize of £5 to Cambridge, while Dean, we suppose, considered himself "highly commended." Next year, however, the indefatigable Dean again competed with the inseparable Cambridge, and with such success that he was awarded the first prize of £10, his rival carrying off the second prize of £5.

Fig. 4.—Dean's Portable Engine, 1844.

An idea seems to have been prevalent at this time that the simpler in character the engine the better it was adapted for use on the farm; and a year or two later this found expression in a statement by the engineer of the Royal Agricultural Society to the effect that "no engine intended for farm purposes should have a superfluous part." A degree of simplicity, almost amounting to rudeness, was insisted upon in the endeavour to obtain an engine suited to the comprehension of the farm labourer; and it is very probable that many obvious improvements were sacrificed by the makers with the view of commending their particular engines to the favourable notice of the judges, as being "devoid of complication." Economy of fuel appears to have been quite a secondary matter, safety being put forward as the first consideration; and it is probable that this attitude on the Society's part did much to retard improvement.

About this time Messrs Clayton & Shuttleworth made what

appears to have been the first portable engine with horizontal cylinders on the top of the boiler, the crankshaft being geared by a wheel and pinion to a second shaft upon which the flywheel was fixed, making three revolutions for one of the crank, the whole being fixed upon a wooden frame. Up to this time there seems to have been a remarkable prejudice against horizontal cylinders, Cambridge, for example, going to the trouble, for the sake of getting a vertical cylinder in connection with a horizontal

Fig. 5.—Cambridge's Portable Engine, 1847.

boiler, of sinking the cylinder bodily into the latter, and adopting an arrangement at that time used in marine practice, of two piston rods carrying between them a bent crosshead projecting downwards into a deeply-recessed top cylinder cover.

At the Newcastle-on-Tyne meeting in 1846, Mr Cambridge was the sole exhibitor of a portable engine, the locality being probably too remote for other makers to incur the expense of sending their engines for exhibition. This engine seems to have been made from the following recipe: "Take a small pillar letter-box, bore out the lower part to fit a piston, and apply a slide-valve and bottom cover; put a crankshaft through near the closed top with the crank inside; couple to the piston by a connecting rod. Fill the space above the piston, which is single acting, with exhaust steam, and immerse the whole in the boiler as far as it will go. Boil till further notice." Had Mr Cambridge only made this engine with two or three cylinders instead of one, it would have borne a remarkable likeness to some modern high-speed engines, and some existing patents would have been anticipated.

At the Northampton show in 1847, seven makers competed for a prize of £50. The award fell to Cambridge, who certainly deserved it if only for his perseverance. His engine now assumed another of its Protean forms, and is shown in Fig. 5, which was reproduced from a photograph taken at the Kilburn show of 1879, at which this identical engine was exhibited. The boiler is shown in Fig. 6, and consisted of a plain cylinder, with an oval flue running through it, divided from the bridge onwards into three longitudinal compartments. The products of combustion traversed this distance three times before escaping to the chimney, the lower part of which was immersed in a tank which formed the feed water heater.

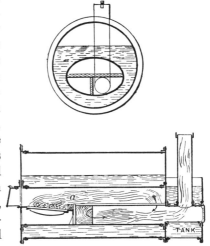

Fig. 6.—Cambridge's Boiler, 1847.

It appears that during the trial this engine was worked at 68 lbs. pressure, a very high one at that time, and at a speed of no less than 250 revolutions per minute. Mr Cambridge was thus somewhat in advance of his time. As usual, objections and protests were lodged, but the engine was tested again before a special committee at a more moderate speed and lower pressure, with the result that the award of the judges was confirmed. Unfortunately, however, the engine now before us does not appear to have come out by any means well, as regards either fuel consumption or first cost, as compared with Trevithick's engine of 1811; but the tests in either case available for comparison are not perhaps as precise as some which have taken place in later years. For example, the competitive trials at the Northampton meeting were conducted thus:—The engines were all placed in a row, each one with its thrashing machine coupled up to it. A certain number of sheaves to be thrashed was given out to each one; at a given signal the series was started, and the first to finish was declared the winner. The duration of the trial was only eight or ten minutes.

Cambridge was again closely followed by his old rival Dean (now Ryland & Dean). Messrs Hornsby, of Grantham, were also among the exhibitors; and there was an engine there by Mr Ogg, of Northampton, which deserves notice in passing. It was a double-cylinder engine of 5 H.P., and had a locomotive type boiler, feed water heater, expansion valves, and other modern improvements, which make it matter for wonder that it did not take a better place in the experiments. It came out fourth. Its weight was about $2\frac{1}{2}$ tons.

The use of double-cylinder engines was shortly afterwards expressly discouraged by the Society, on the ground that "two cylinders are not necessary for agricultural purposes; they make the engine more expensive; there are more parts to be kept in repair, and more attention is required to work them,"—all of which is profound truth to this day.

In Fig. 7 is shown an engine exhibited by Messrs Hornsby, of Grantham, at the York meeting in 1848, where it was successful in carrying off the first prize of £50. The engine was lent by its owner as a contribution to the "Museum of Antiquities" at Kilburn in 1879. It will be noticed that the firebox is circular; the piston rod is prolonged, and passes through a stem guide, with a long forked connecting rod. The steam ports were short, one at each end of the cylinder face, with the exhaust port near to one of them. A hollow slide-valve, nearly as long as the cylinder, allowed the exhaust steam to pass through it from the further port into the exhaust opening. The cylinder was 10 in. in diameter, and of 14 in. stroke, and the engine is recorded as having run against a 6 H.P. load, with a consumption of 7 lbs. of coal per horse-power per hour. This, however, is probably an error. The same, or a similar, engine consumed 14·2 lbs. of coal per horse-power at the Norwich trials in the following year. Messrs Hornsby subsequently made this class of engine with an oval trunk working through the front cylinder cover, and one of these the writer, as a boy, assisted to place upon a

Fig. 7.—Hornsby's Portable Engine, 1848.

new boiler at Messrs Hornsby's works at Grantham, in the year 1872.

The judges at the Exeter show in 1850 mention that Messrs Clayton & Shuttleworth's engine, there exhibited, had the steam pipe carried through the exhaust pipe, the latter itself passing through the boiler. As we shall have something to say about inside exhaust pipes shortly, we shall not comment further upon this remarkable arrangement of a doubly steam-jacketed exhaust pipe.

Very shortly after this we find nearly all the principal makers turning to the modern type of portable engine, with the ordinary locomotive type boiler and a single horizontal cylinder upon the

Fig. 8.—Bach & Co.'s Portable Engine, 1850.

top of it. There is a statement in *Engineering*, in its account of the Kilburn show, that the first maker of engines of the now universally accepted pattern was Mr Richard Bach, of Birmingham, (Mr Bach's engine is shown in Fig. 8, and is quite a creditable machine, considering its early date). There were, of course, exceptions—for instance, the "steeple engine" of Messrs Tuxford, which they exhibited at Exeter for the first time in 1850, and continued to manufacture for many years afterwards. The engine itself shown in Figs. 9 and 10, is contained in an iron housing at the end of the boiler furthest from the firebox. The cylinder is placed vertically, the piston rod carrying a short beam or crosshead, from the ends of which a pair of rods descend, having guide-blocks

Fig. 9.—Tuxford's Portable Agricultural Engine, 1850.

attached to their lower ends, sliding in guides formed in the sides of the cylinder itself. From these blocks a couple of connecting rods rise to a very wide crank placed just above the cylinder. The boiler has a flat flue leading to an internal smoke-box at the engine end of the boiler; from this, tubes return to the chimney over the firebox. In later engines, by the same firm, this arrangement was reversed; the tubes passed from the firebox in the ordinary way, and a flat flue over them returned to the chimney.

In 1851, at the Crystal Palace trials in Hyde Park, in connection with the great International Exhibition, eleven makers entered engines for competition, the first prize being awarded to Messrs Hornsby, of Grantham, whose engine, Fig. 11, was officially

Fig. 10.—Tuxford's Engine (End View), 1850.

described as having "a horizontal cylinder, fitted centrally in the steam-dome over the firebox; the boiler covered with dry hair-felt and wood, and the feed water heated in the smoke-box"—an excellent specification, we may say. The following table, extracted from the Jurors' Report, gives the comparative efficiencies of the engines reported upon:—

EVOLUTION 17

COAL PER HORSE-POWER PER HOUR.

Hornsby	- -	6·73 lbs.
Tuxford	- -	7·46 ,,
Clayton	- -	6·63 ,,
Garrett	- -	8·65 ,,
Barrett	- -	9·20 ,,
Tuxford	- -	10·85 ,,
(4 H.P. oscillating cylinder)		
Cabron	- -	12·48 lbs.
Burrell	- -	13·10 ,,
Butlin	- -	14·71 ,,
Hensman	- -	18·75 ,,
Roe	- -	25·8 ,,

Fig. 11.—Hornsby's Portable Engine, 1851.

It is probable that the performance of Messrs Roe's engine, in the way of coal consumption, has never been surpassed by any other maker at a Royal show trial, but this record was completely eclipsed by the same firm at the Lewes meeting in the following year, where their engine is stated to have achieved the colossal result of 93·9 lbs. per horse-power per hour.

Messrs Garrett's engine, the fourth on the list, is shown in Fig. 12. It is officially described as a "light, strong, portable engine with an external horizontal cylinder."

Messrs Ransomes, Sims, & Jefferies, Ltd., have kindly sent us a very old photograph, showing an engine made by them in 1851, though not, we believe, exhibited. This engine, which we reproduce with great pleasure, Fig. 13, has the peculiarity of being mounted on springs, and is still (1911) doing useful work in the county of its origin. Messrs Clayton & Shuttleworth, who, at the Norwich show of 1849, had taken the second prize, exhibited for the first time, at Gloucester in 1853, their inside-cylinder engine, Fig. 14. As will be seen, the cylinder is enclosed within an upward extension of the smoke-box. The cylinder was also steam-jacketed, and these improvements were credited with the reduction of the coal per horse-power to 4·32 lbs. per hour, the lowest as yet recorded. This was, no doubt, very largely also due to the excellent proportions of the boiler. This inside-cylinder arrangement was considered at the time a most important improvement, as doubtless it was, though discarded many years since.

Fig. 12.—Garrett's Portable Engine (Firebox End), 1851.

No further trials were held by the Society until the year 1858, when, at their Chester meeting, preparations were made for a more elaborate series of trials than had been before attempted, and under much more stringent conditions, especially in the matter of simplicity of construction, and facility in taking to pieces. For instance, the feed pumps were not to have more than two valves, both to be perfectly accessible. How the accessibility of the second valve would be promoted by the absence of a boiler check-valve must be left to the imagination of the reader.

In spite of these regulations, but probably on account of more accurate testing, the coal consumption at these trials does not show such a good average result as was achieved by the same makers three years before. The practice of preparing special

Fig. 13.—Ransomes' Portable Engine, mounted on Springs, 1851.

engines, or "racers," for trials, was also strongly condemned in the report of the judges at this meeting.

The type of portable-engine boiler shown in Fig. 15 was patented in 1861 by Messrs Robey & Scott, of Lincoln. It will be seen that the water-space is carried round under the firebars, and a highly successful boiler, both mechanically and commercially,

20 THE PORTABLE STEAM ENGINE

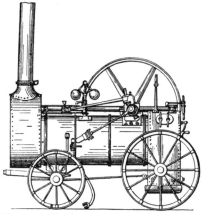

Fig. 14.—Clayton & Shuttleworth's Inside Cylinder Portable Engine, 1853.

this turned out to be. Its only disadvantages were additional weight and increased first cost, but these were more than counter-balanced by its increased steaming power, and the exceptional facilities afforded for the deposit of sludge and dirt in the bottom space, whence they could be easily removed. It is worthy of remark that this identical construction was, many years later, reinvented by the late Mr F. W. Webb, of Crewe, and applied to the boilers of his compound locomotives. Messrs Robey, it may be remarked, never entered their engines for trial at the Royal Agricultural Society's meetings.

Fig. 15.—Robey & Scott's Portable Engine, with Water-bottom Firebox, 1861.

Further trials were carried out by the Society, at their Worcester show in 1863, the best result being credited to Messrs Tuxford for their 8-H.P. engine, fitted with an expansion-valve, and consuming 3·59 lbs. of coal per brake horse-power per hour.

The next great trials were those at Cardiff in 1872, where the

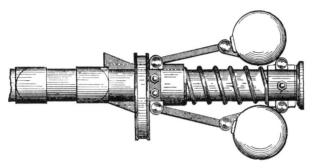

Fig. 16.—Robey's Crankshaft Governor, 1869.

honours were divided between Messrs Clayton & Shuttleworth and the Reading Iron Works Co., for engines reported to consume 2·71 and 2·79 lbs. respectively. In 1869 an engine was exhibited at the Smithfield Club Cattle Show, fitted with the crankshaft governor shown in Figs. 16 and 17, acting directly, by what is known as the wedge-motion upon the slide-valve eccentric, thus forming an automatic expansion gear. This arrangement was patented in the same year by Robert Robey and John Richardson. After a careful search, we are unable to discover any trace of automatic expansion gear being actually applied to the portable engine prior to this, its use having been confined to stationary engines of large power.

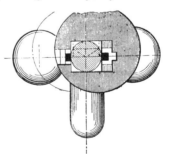

Fig. 17.—End View of the Robey Governor, 1869.

The history of the portable engine may be considered as ending at the Cardiff trials of 1872, inasmuch as all further improvements now extant are matters of detail merely. No new principle of construction, so far as we know, has been evolved or sought after, and the attention of manufacturers has been mainly directed to strengthening, improving, and generally, under the wholesome stress of competition, seeing who could produce the best possible article for the money. It affords an example of what biologists call "reversion to type" to remark how such variations, as, for example, Tuxford's steeple engine, the inside cylinders of Clayton and of Hornsby, and Robey's closed-bottom firebox, have all disappeared, leaving now little to distinguish the engines of one maker from those of another.

Briefly reviewing the improvements in matters of detail during the last thirty years, we should be inclined to put in the first place the improved system upon which the boiler is constructed. A good many years ago, it was tolerably easy to turn out a fairly good engine, but the art and practice of boiler-making had not then developed itself. Those of us who are of middle age have had the opportunity of seeing a marvellous change in this respect, due very largely to the immense improvement in the material worked upon, and the possibility of procuring steel plates of reliable quality at a price which has practically driven iron out of the field.

With this beautifully ductile, almost amiable, material to work upon, with the flanging press, with the cleverest drilling machines, with all manner of self-acting tools for planing, turning, and boring all plates, with the system of annealing the plates after being worked in the fire, and, finally, with the hydraulic riveter, the building of a boiler may now be considered the assembling together of a number of accurately fitted parts, rather than a system of forcible combination of pieces of plate by the aid of various persuasive instruments known as drifts and cramps. It is, however, only fair to say that for years before the practice of punching the rivet holes in boiler plates was abolished in favour of drilled holes, the workmen had become rather expert at it, and a set of boiler plates punched from templates while straight and afterwards bent into shape would "come together" with a marvellous degree of accuracy all things considered. But in modern portable engine practice iron boiler plates and punched holes are things of the past, and however good the engine may be, there is the knowledge that it is mounted upon a boiler just as accurately and as carefully constructed as itself.

Another valuable, though less obvious, improvement of recent years has been the greater regard paid to the accessibility of the engine parts. It is no longer considered a desirable thing to pack away everything possible inside the boiler or inside the cylinder. The very early makers, in common with the latest present-day practice, had the virtue of putting everything where it could be seen and got at. But, between these two eras came a period of neatness and snugness. The exhaust pipe passed through the steam space of the boiler, the stop-valve, throttle-valve, and safety-valve were within or upon the cylinder, and a general effect of sleekness and smoothness was evident.

Inside exhaust pipes absorb and convey away up the chimney a considerable amount of heat, are difficult to get at, and worse to get out, for purposes of repair, and there is always the possibility of unseen and unsuspected leakage of live steam into the exhaust, owing to decay of the pipe or a broken joint. The steam inlet was also underneath the cylinder, affording another possibility of leakage in an inaccessible place.

At that time the cylinder casting of a portable engine was a trophy of the art of the moulder. The separate liner, or working barrel, now universally used, was at that time unthought of, and the double walls, forming the steam jacket, with its inlet passages; the steam and exhaust ports, and their inlet and outlet pipes; the throttle and stop-valve seats, the safety-valve seats, and various other details were all cast in one piece with the cylinder and steam chest. In a double-cylinder engine, all this complication was doubled, still in a single casting; and the number of cores in a cylinder of this kind made it a matter for wonder that a successful casting could ever be hoped for.

See how these are remedied in the modern engines in the next chapter, and referring back to Fig. 5, notice that Cambridge's engine of 1847 is fully up-to-date in this respect. The stop-valve is outside the cylinder, and the exhaust pipe runs along the boiler, just as in the more recent engines—outside. It can, if necessary, be removed in a few minutes, and when worn out, does not allow live steam to pass into it, and escape unsuspected, up the chimney. In the modern engines, presently to be described, you will also observe that the plummer-blocks, or main crankshaft bearings, are connected to the cylinders by stay-rods, the former being allowed to slide upon their brackets to allow for the expansion of the boiler while getting up steam.

The much-discussed subject of stay-rods claims a moment's attention. It is evident, at first sight, that the conditions differ very much in a portable engine from any other in which the engine is not erected upon a boiler liable to continual alteration in length by expansion and contraction. But, far worse than no stay-rod at all was the practice, adopted up to only a few years

Fig. 18.—Friction Diagrams from Engine with Throttling Governor.

ago by makers who ought to have known better, of connecting the cylinder to fixed or non-sliding plummer-blocks, presumably with the idea of preventing the expansion of the boiler by main force, the result being that an enormous stress, compared with which the ordinary pull-and-thrust of the engine is a mere nothing, is put upon the boiler and all the fixed parts of the engine. In fact

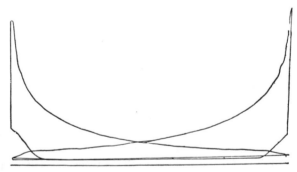

Fig. 19.—Friction Diagram from Engine with Automatic Expansion Governor.

it affords a modern solution of a question much discussed by ancient philosophers, as to what would be the effect of an irresistible force acting upon an immovable body? At the present day the effect is generally to cause leakage of the boiler, and to develop cracks in the cylinder where the stay-rods are attached.

Another passing phase or fashion in portable engines has been the application of automatic expansion gear, an early example of

EVOLUTION 25

which is shown in Figs. 16 and 17. It may perhaps be worth while to discuss briefly the theoretical and the practical sides of this question. It is well known, of course, that the number of revolutions per minute of an engine is kept nearly constant for all conditions of load by the action of the centrifugal governor, which checks the tendency to "run away" when a part or the whole of the load is removed, by restricting the supply of steam, and *vice versa*. This may be done in either of two ways. Firstly, the governor may (as in the vast majority of portable engines now manufactured) act upon a throttle-valve in the steam inlet, and by giving either a little less or a little more opening correspondingly decrease or increase the steam pressure in the valve chest. Secondly, the governor, of a larger and more powerful type, so

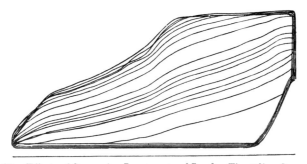

Fig. 20.—Effects of Successive Increments of Load. Throttling Governor.

adjusts the travel of the outer, or expansion, slide-valve as to regulate, not the pressure, but the quantity of steam supplied, the valve chest being in free communication with the boiler always, while the engine is at work.

We have here two sets of indicator diagrams, taken by the writer from two Robey portable engines, identical in all respects, save that the second engine was fitted with the makers' automatic expansion gear. Figs. 18 and 19 are friction diagrams, that is to say, the load has been removed, and the engine is running light. Figs. 20 and 21 show respectively the effects of successive increments of load, as recorded by the indicator, for the two systems of speed-regulating. It will be seen at once that the *pressure* varies in Figs. 18 and 20; and that in Figs. 19 and 21 the pressure (shown by the height of the diagram) is constant, but that the *quantity* is varied by altering the point in the stroke at which the steam is cut off.

The argument in favour of the automatic system is that steam at virtually boiler pressure is admitted to the piston, and is cut off, dead, at the point corresponding to the load at the moment, thereby securing the full benefit of expansion during the remainder of the stroke, up to the opening of the exhaust. This is of necessity accompanied by stresses upon the working parts of a very different character from those imposed by the alternative system of governing by throttle. In the latter case a lower pressure is admitted through a longer period, resulting in a smoother working, and less frequent necessity for repairs and adjustments. We think that now (1911) it is pretty well established that for the usual purposes of a portable engine, the greater

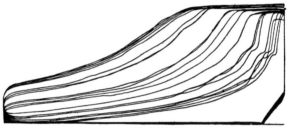

Fig. 21.—Effects of Successive Increments of Load. Automatic Expansion Governor.

simplicity and lower first cost of the throttling system make it, on the whole, the more desirable investment, at any rate for non-compound engines of ordinary sizes.

Compound portable engines of large sizes are now a standard article of manufacture; while an offshoot, in the form of semi-portable engines, has within the last few years developed, principally on the Continent, to an extent which could hardly have been anticipated, engines up to several hundred horse-power now being obtainable of this type.

We shall now proceed to examine the standard types and classes in which portable engines are now manufactured; and we think our readers will agree that in design and construction they are worthy examples of engineering progress in this present year of 1911.

CHAPTER II

THE PORTABLE STEAM ENGINE OF TO-DAY

THE characteristic essential of the portable engine is, naturally, its portability; and as we shall see in a moment, this condition is not altogether satisfied by the fact that it is mounted upon four wheels like any other heavy vehicle.

The motto of the portable engine is *Ubique*, and, as we have already remarked, its work lies, not seldom, at the very outposts of civilisation, so that instead of the engine following the roads, the road may very literally be said to follow the engine. Forests have to be cleared, swamps drained, bridges of stone or timber constructed, and rocks blasted and broken up for road metal before vehicles can travel in a civilised and decorous manner.

Conveyance upon its own wheels, as it nears the end of its journey, is sometimes impossible, and there is nothing for it but to dismount the engine from the boiler, and carry the separate parts bodily, by sheer force of human labour. Stripped of everything which can be taken from it, the boiler has to be floated across streams, lashed to timbers and supported by empty casks to give it buoyancy. In passing, it may be useful to note that small portable engine boilers are too heavy in proportion to their volume to float without assistance, even when all the openings are sealed up. Before now, a boiler of this kind has been built up solid with timber into a cylindrical form, and rolled up a mountain side, or lowered down a precipice in the manner adopted by the brewers' drayman. In more serious cases it sometimes happens that the entire machine has to be transported in small pieces, each forming a load for one or two men. Here, however, the ordinary portable engine because of its heavy unit, the boiler, fails utterly, and recourse must be had to a water-tube boiler, which, conveyed in single tubes, can be taken up a ladder if necessary. But this is by the way.

One of the principal modern improvements in the direction of portability is the detachable engine. Under the old system the cylinder casting is curved to fit the boiler, and is secured to it by bolts passing through the boiler plates; while the saddle, or casting comprising the two plummer-blocks for carrying the crankshaft, is also curved to suit the boiler, and attached by more bolts. This method of construction strikes us at once as offering at least two disadvantages—one, the obvious liability of leakage through any one of the sixteen bolt holes passing through the shell of the boiler, and the other the fact that the cylinder is, so to speak, specialised, in that it is individually fitted to the curve of the boiler, and therefore cannot be replaced, if broken or worn out, without the services of a skilled portable engine erector—a trade to itself, whose professors are not as readily available at Tientsin or Timbuctoo as in the workshops of Lincoln or Gainsborough.

The same objections apply, of course, but in an even greater degree, to the cast-iron saddle, with the additional disadvantage that the latter is particularly liable to breakage when the engine is stripped for travelling over mountain paths.

There is one more accusation to be brought against the cast-iron saddle while we are on the subject, and that is the obvious want of a direct connection between itself and the cylinder, to take up the stresses induced by the pull-and-thrust of the piston upon the crank. As it is, these stresses are conveyed through the boiler—not a desirable thing in itself, although, in the smaller sizes, the boiler is strong enough to form the frame of the engine without injury. The chief objection to this arrangement is the possibility of its giving rise to leakage at the bolt holes already mentioned, unsuspected and unseen until the removal of the lagging (or sheathing of non-conducting material which covers the boiler) discloses an unwelcome spectacle—the wasting away of the boiler plate in the wake of some one or more of the bolt holes—a defect not easily remedied.

Owing to the expansion and contraction of the boiler as it is alternately heated and cooled, anything in the nature of a rigid connection between cylinder and saddle is out of the question, as it would give rise to stresses immeasurably greater than those it was intended to absorb. This brings us to the method now almost universally adopted for rigidly connecting the cylinder and crankshaft bearings while allowing for the free expansion of the boiler—the sliding plummer-block. Upon the steel-plate brackets riveted—not bolted—to the boiler barrel, cast-iron fitting pieces are fixed, upon which the plummer-blocks are free to slide longitudinally by means of a dovetailed rib, or its equivalent.

A steel tie-rod, or stay, connects each plummer-block with its own side of the cylinder, the latter being rigidly fixed to corresponding brackets riveted to the firebox casing. This is the general principle of the detachable engine, carried out in differing detail in the engines we shall illustrate.

This allows of a complete and independent horizontal steam engine being erected with all its parts separately upon a surface plate in the workshop, with all its bolt holes drilled to template, without reference to the particular boiler it may be connected up to. From this it follows that, in the event of serious damage to the boiler, as for example the not unknown case of a fire being lighted without water in the boiler, a new boiler could be sent out with the certainty of fitting correctly to the existing engine. Conversely, any details required for the repair of the engine, or the complete engine itself, can be ordered without fear of any difficulty in its erection upon the boiler.

If circumstances require it, the engine and the boiler may be dissociated and put to work in separate rooms, or placed side by side. Only those who have been in charge of steam-driven machinery, in situations far removed from the possibility of obtaining skilled mechanical labour, can fully appreciate the marvellous adaptability of the portable steam engine in its modern form, as illustrated in the pages which follow. True economy in steam consumption is only attained by making the working parts strong and stiff enough to withstand the stresses induced by an early cut-off in the cylinder with a high initial steam pressure—otherwise the benefits of expansive working are more than neutralised by endless trouble caused through bending stresses in the engine parts, leading to hot bearings and other difficulties.

The boiler of the portable engine, always, of course, of the locomotive type, is nowadays invariably of steel, and is, as a rule, an admirable specimen of good design and accurate workmanship. So far as we have been able to ascertain, there are only three variations from the normal construction in general use, which we shall consider in due course. Neither of these, however, affects the general principle of the boiler. The material is, as we have said, steel, but as that convenient appellation covers a very wide field, it will be necessary to define the quality and grade of the steel which it is desirable should be employed in a first-class locomotive type boiler.

Boiler shell plates should have a tensile strength of not less than 26 or more than 30 tons per square inch of section. A sample of Siemens-Martin steel, of thoroughly satisfactory character, gave the following results:—Ultimate tensile strength, 27·9 tons; elongation in 8 in., 26 per cent.; contraction of fractured area, 47 per cent. A strip of this plate, heated to a cherry-red, and plunged into water at 80° Fahr., may be doubled over flat upon itself without any indication of cracking. The drifting test shows that a $\frac{5}{8}$-in. hole may be enlarged, cold, to $1\frac{5}{8}$ in. diameter without distress of the plate.

An analysis of the same piece gave:—Carbon, ·16 per cent.: silicon, ·018 per cent.; sulphur, ·035 per cent.; phosphorus, ·066 per cent.; manganese, ·492 per cent. A similar analysis of firebox plate gave respectively: ·16, ·022, ·032, ·05, and ·479 per cent. This plate showed 25½ tons tensile strength, 28 per cent. elongation in 8 in., and 48 per cent. reduction of fractured area.

In rivet steel the tenacity is 26 tons, but the ductility is greater, the elongation in 8 in. being 31 per cent., with a contraction at fracture of 66 per cent. Rivets should admit of being doubled over flat, cold, and the head should allow of being flattened out, hot, to $\frac{1}{8}$ in. thickness without cracking at the edges.

The flanging of the plates should be done entirely by hydraulic pressure; and all rivet holes and stay holes should be drilled in position, the plates being afterwards separated, and the small burrs produced by the drilling (which would interfere with the close contact of the plates) removed. The edges of all the plates in the boiler should be machined. Longitudinal seams should be butt-jointed and double riveted. Hydraulic or machine riveting should be employed throughout, a variety of special portable riveting machines now being available, capable of dealing with difficult positions.

Plates which have been locally heated for any purpose should not be put into the boiler until they have been annealed, which has the effect of setting the plate at rest, or, in other words, of eliminating all local stresses induced by partial heating. The manhole, of course, should be adequately reinforced by a strengthening plate riveted round it, and the manhole cover should be of pressed steel, preferably of that kind which has a flat joint, rather than one which follows the curve of the boiler. As a rule, however, in dealing with any firm of good repute there is little need to go beyond their own standard of material and workmanship, which will always be found to fill the ordinary requirements of any boiler insurance association.

The rules for the strength and staying of locomotive type boilers are too complex for discussion here,* but it may be taken for granted that if the boiler stands an hydraulic test of 100 lbs. per square inch above the ordinary working pressure, whatever that may be, there will be little fear of faulty workmanship or bad design escaping detection.

In the general construction there is very little difference between the boilers manufactured by the leading makers. We have mentioned three specialities which form almost the only exceptions to the ordinary design; all consisting of improved methods of supporting the roof of the firebox. It may be worth while to explain the ordinary method of doing this, and the need for improvement which undoubtedly existed. Few people, other than those interested in the design or manufacture of locomotive type boilers, realise the pressure which has to be sustained by the firebox top of quite a small portable engine boiler.

Take, for example, the firebox of an engine having a single 10-in. cylinder, usually rated as 8 H.P. nominal, which would have about 5·75 sq. ft. of grate area. Adding the 25 per cent. extra for wood burning, we have 7·19 sq. ft. as the area of the roof plate. At 100 lbs. steam pressure per square inch, or 14,400 lbs. per square foot, there would be 103,536 lbs., or more than 46 tons, as the total downward pressure to be resisted all the time the boiler is at work.

How is this considerable pressure, which would crush in a flat roof plate instantly, to be provided against? The usual plan

* See the author's paper, "On the Structural Strength of Loco Type Boilers" (London, Feilden Publishing Co., Ltd.).

32   THE PORTABLE STEAM ENGINE

is to provide a sufficient number of roof girders, resting at their ends partly upon the rounded corners of the vertical end plates of the firebox, and partly upon those portions of the roof plate itself immediately contiguous, the latter being hung up to the girders by stay-bolts, at a pitch (or distance apart from centre to

Fig. 22.—Ordinary Method of Supporting Firebox Roof Plate.

centre) carefully proportioned to the load each bolt is capable of carrying safely.

The immediately visible defects of this arrangement (Fig. 22), which, nevertheless, is the one adopted in the vast majority of portable engine boilers, are, first, that the whole of the stress is transmitted to the end plates of the firebox, one of which—the tube plate—is so much weakened by the large area cut out for the tube holes, that a much greater stress than is desirable is imposed upon the solid portions of the plate between the holes. To make up for this by thickening the tube plate so that the metal is not unduly stressed, would, if carried out completely, result in a plate of such thickness that it would become overheated.

The second obvious disadvantage of the ordinary roof girder system is, that the space between the girders and the plate—usually not more than about $1\frac{1}{4}$ in.—owing to the difficulty of access from the manhole, sooner or later becomes filled up solid with scale or deposit, thus causing overheating of the roof plate. It should, perhaps, be explained that the plates of a firebox, unless in actual contact with the water, soon reach a temperature at which the strength of the plate is seriously impaired.

We may, in passing, allude to a method now seldom, if ever, employed in portable engine boilers, by which the roof girders are eliminated altogether. This is the system of sling stays, in which the roof plate of the firebox is curved (or cambered, to use the technical phrase), and the stay-bolts are carried right up to and screwed through the arch plate of the external firebox. This method introduces two incidental disadvantages. The first is that the stay-bolts, connecting the cambered plate of the inner firebox with the semicircular plate of the external one, have to be screwed through both plates at an angle which is a compromise between the radial directions of two differing curves, with the consequence that they pass through both at an unfavourable angle. The second disadvantage of this arrangement is that there is no elasticity about it. When the inner firebox expands upwards through the heat, which occurs before the external box becomes heated to the same degree, an enormous compressive strain is exerted upon these sling stays tending to force them through, and loosen them in, the screwed holes through which they pass.

Messrs Richard Garrett & Sons, Ltd., realised the difficulties and disadvantages of the then universal system of roof bars, so far back as the year 1876, since which date everyone of the many thousands of loco-type boilers turned out at the Leiston works have been fitted with the type of firebox shown in Fig. 23.

The makers state that this firebox is peculiarly free from incrustation and the collection of mud on the crown, owing to the absence of the usual stays. The natural expansion and contraction are alone sufficient to prevent incrustation, as any thin coating of mud or scale which may collect is immediately broken on the next occasion of any change of temperature (with the consequent alteration in area and form of crown plate), and carried away by the violent ebullition which takes place above this—the hottest—

Fig. 23.—The Garrett Corrugated Roof Plate.

portion of the firebox, to settle ultimately in a specially designed mud pocket from which it is easily removed. In addition to these two good points of extreme durability and strength, the firebox has also the advantage of being the most efficient, as, owing to its form, it contains more heating surface than any other firebox of similar grate area. The pocket between the corrugations may really be described as the lower and more efficient half of a water tube, situate in the midst of the hottest gases of combustion.

THE ENGINE OF TO-DAY   35

Messrs Robey's solution of the problem is shown in Fig. 24, and consists in principle of transverse roof girders, carried (at a height above the firebox crown plate of some 6 or 8 in. to allow of free access for cleaning) by two brackets, or specially formed

Fig. 24.—The Robey Patent Roof Girders.*

steel angles, riveted to the shell of the firebox. As these girders are evidently free to lift with the expansion of the firebox, elasticity is secured, while the entire weight (the 46 tons, for example, already mentioned) is carried by the external, or firebox-shell, plate; thus relieving the foundation ring (or frame connecting

* Specification of John Richardson and William Dyson Wansbrough, No. 6954 of 1900.

the inner and outer fireboxes) from the liability to leakage, always present when it has to take the pressure due to the whole area of the firebox. This system, which has been employed in all boilers manufactured by Messrs Robey since its introduction, has greatly increased the life of the firebox, and has almost eliminated a very common source of trouble, the breakage of the screwed side stays.

Fig. 25.—Marshall's Patent Firebox.

The third method of protecting the crown plate is shown in Fig. 25, and is the speciality of Messrs Marshall, Sons & Co., Ltd. of Gainsborough, who state that they are convinced that this is the strongest and safest crown plate of any of the various forms of corrugated tops which are on the market. The corrugations spring from opposite corners of the firebox, crossing diagonally in the middle, forming an exceptionally strong truss to the crown plate. These corrugations are obtained with the least possible distress to the material of the firebox, and their strength is proved by tests, which have been repeatedly made. The shape of the corrugations enables them to breathe freely, and the absence of cumbrous stays enables the top to be readily cleaned.

From all this it will be seen that there are many things to be considered, even in the supporting of the firebox crown of a small portable engine boiler.

In the construction of the boiler shell, there are only two alternative systems in use, so far as the locomotive type is concerned, each of which has its own special advantages in connection with the portable engine. These are the straight-line, or flush-topped, boiler, and the raised firebox, or saddle boiler. The latter takes its name from the saddle, or flanged junction plate uniting the barrel, or cylindrical part of the boiler, with the firebox shell. As a consequence of this construction, the barrel is some $2\frac{1}{2}$ or 3 in. smaller all round than the arched top of the firebox shell. As, however, there is still room in the barrel to accommodate as many tubes as the breadth of the inner firebox allows of, there is no restriction of the heating surface through the lessened barrel diameter.

This pattern is a favourite one with portable engine builders, as when the crankshaft is fixed at such a height that the crank just clears the barrel, the superior height of the firebox casing is exactly adapted to receive the cylinder; engine and boiler thus affording an example of mutual accommodation. The disadvantage of this system is, naturally, the set-off in the outline of the boiler, giving longitudinal elasticity to the structure. This effect, however, can be, and of course always is, resisted by an adequate provision of longitudinal stays extending from end to end of the boiler. These are usually plain steel tie-rods screwed through one or both the end plates, with nuts inside and outside. The diameter of these tie-rods should be equal throughout to secure uniformity of stress, that is to say the screwed parts should be formed on an enlarged portion of the rod, the difference in the diameters being twice the depth of the screw thread. The use of this boiler is mostly confined to the smaller sizes and lower steam pressures, for which it is in every way admirably adapted.

The straight line boiler has its barrel diameter equal to the width of the firebox shell, and is usually spoken of as the "flush-topped" type. Most of the makers adopt this pattern for their larger sizes, its increased capacity of barrel giving more steam room, and greater facilities for entering and cleaning the boiler. Longitudinal tie-rods are, of course, still necessary, but are mainly for the purpose of staying the end plates, viz., the smoke-box tube plate, and the upper part of the firebox front plate. In some cases, however, these flat surfaces are stayed by "gussets," which are triangular pieces of plate, bracketing the shell and end plates together.

Having now noticed at some length the leading features to be considered in the design and construction of a successful portable engine and boiler, we shall now have the opportunity of seeing how far the requirements we have sketched out are fulfilled by the principal manufacturers in this country, some of whom have turned out, since their establishment, twenty, forty, and even sixty thousand portable engines.

They have put into their work, as it went on, the results of accumulated experience, tending always towards simplicity. The "fads" and eccentricities common thirty years ago have one by one disappeared; and, with very minor differences in detail—as an inspection of our illustrations will show—the present-day portable engine has settled down into a permanent type; the product of independent methods converging upon a central idea.

CHAPTER III

SOME STANDARD TYPES OF SINGLE-CYLINDER ENGINES

By the courtesy of the leading manufacturers, we are enabled to present examples of the latest practice in single-cylinder engines. As will be seen, the notices of the various firms' productions are alphabetically placed; and the claims of each maker are as far as possible given in their own words.

It would be presumption on the part of the author to attempt any criticism where such efforts are manifest upon every hand of a desire to study efficiency and economy rather than first cost.

The names we give here are, without exception, of world-wide celebrity; and, from the writer's own knowledge, their industrial establishments in this country (and in many cases abroad also) are magnificent examples of the highest skill and organisation applied to one end—the production, upon a huge scale, of the articles they have to offer, through the means of every labour-saving appliance which money can purchase or ingenuity devise.

With the view of presenting each firm's specialities upon a common basis for purposes of comparison, the engine we illustrate in each case is understood to be the 8 H.P. (nominal) size of the standard single-cylinder class, exactly as made and sold without any extras whatever unless specially indicated. Nominal horsepower, it may be noted in passing, is simply a catalogue reference, and is in some cases discarded in favour of a number or letter.

Compound engines, and engines of special classes, will be dealt with separately, under another heading.

Messrs Brown & May, Ltd.

Messrs Brown & May, Ltd., of Devizes, Wilts, are an old-established firm, having celebrated their jubilee in the year 1904.

Fig. 26.—Messrs Brown & May's Portable Engine.

Their standard 8 N.H.P. engine is shown in Fig. 26, and has a cylinder of 10 in. diameter by 14 in. stroke, giving out at 120 revolutions per minute 24 to 28 brake horse-power (B.H.P.). Features in Messrs Brown & May's engines are their patent grease-separator, their patent steam-blast tube cleaner, and their patent feed water heater, all of which are fitted without extra cost to engines of 4 N.H.P. and upwards. The governor is of the Pickering type (original make) and the crosshead guides are cylindrical.

The cylinder and crankshaft bearings are bolted to steel seatings riveted to the boiler, and are connected by steel stay-rods, the plummer-blocks having a sliding attachment to the brackets to allow for the expansion of the boiler. The working pressure is 100 lbs., at which they are readily insured by the leading boiler insurance companies.

MESSRS CLAYTON & SHUTTLEWORTH, LTD.

To Messrs Clayton & Shuttleworth, Ltd., of Lincoln, is due, we believe, the credit of introducing the present universally adopted type of portable engine upon a commercial scale. Their standard engine is shown in two views, Figs. 27 and 28.

One of the characteristic features of this firm's engines is the flush-topped or straight-line boiler, an incidental advantage of which is that the lagging (or covering of wood and sheet iron) of the boiler is extended so as to cover the upper part of the firebox.

Messrs Clayton & Shuttleworth's cylinder is shown in Fig. 29. As will be seen, it is effectively steam-jacketed, the steam passing through the jacket on its way to the starting valve, the latter being a sliding shutter or valve, worked by a long hand lever. The steam is admitted to the cylinder through the small elbow pipe shown, which forms the only steam joint with the boiler, and can be removed without disturbing any other part. The construction of the cylinder with all its details is so clearly shown in Fig. 29, that we need not dwell further upon it. The planed cast-iron saddle to which the cylinder is secured is bolted to the boiler, as also are the two cast-iron crankshaft brackets. Messrs Clayton & Shuttleworth have not yet adopted the principle of forming these details in steel, and riveting them to the boiler.

Fig. 27.—Messrs Clayton & Shuttleworth's Portable Engine (Pump Side).

STANDARD TYPES

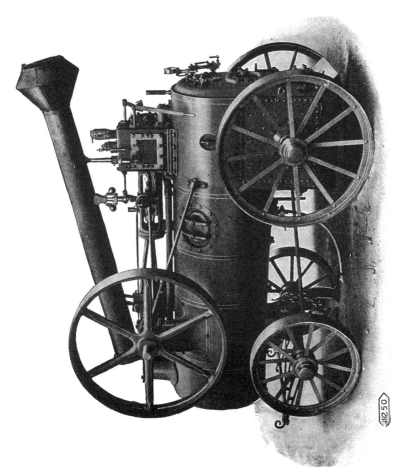

Fig. 28.—Messrs Clayton & Shuttleworth's Portable Engine (Flywheel Side).

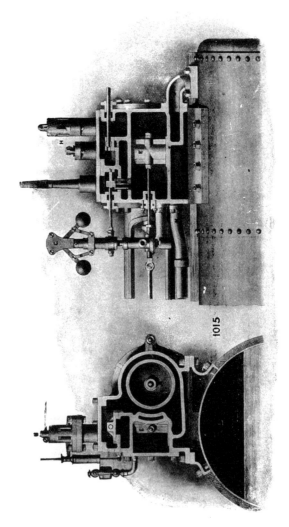

Fig. 29.—Messrs Clayton & Shuttleworth's Cylinder (Sections).

The crankshaft plummer-blocks are of a new and improved design, Fig. 30, being fitted with four-part gun-metal bearings which are adjustable for wear in each direction. Both plummer-blocks are rigidly stayed by tie-rods; the right-hand one in the usual way to a lug on the cylinder, the left-hand one to a point on the firebox shell, a method, we believe, peculiar to Messrs Clayton. The expansion of the boiler, and consequent movement of the brackets under the plummer-blocks, is provided for by a sliding attachment between them.

Fig. 30.—Messrs Clayton & Shuttleworth's Plummer-Block.

The adjustable eccentric, for varying the point of cut-off in the cylinder, is shown in Fig. 31. It will be seen that the eccentric itself is not directly attached to the crankshaft, but is clamped to a fixed circular disc, across which it can be moved to the indications of the pointer, and tightened up to give the period of admission required. The path of the eccentric, in its movement across the disc, is at such an angle with the crank that a uniform lead is maintained whatever the cut-off may be. The marking of the points of cut-off is duplicated, so that variable

46   THE PORTABLE STEAM ENGINE

expansion is obtainable in whichever direction the engine is required to run.

These engines may be fitted at pleasure either with the cross-armed spring-loaded governor shown in Fig. 29, or with the well-known Pickering governor illustrated in Fig. 32.

It should be noted that this engine is fitted with the now very generally adopted trunk, or circular, guides for the crosshead, the

Fig. 31.—Messrs Clayton & Shuttleworth's Variable Expansion Eccentric.

outer end of the trunk being secured to the boiler by a bolted bracket. Finally, the axles of the road wheels merit a passing notice, those for the hind wheels being attached to the firebox shell by circular flanges bolted on, the front axle being connected to the turn-plate by a species of universal joint, which allows of the requisite movement in the fore-carriage when travelling over an uneven road.

STANDARD TYPES

Fig. 32.—Messrs Clayton & Shuttleworth's Cylinder (showing Pickering Governor).

Fig. 33.—Messrs Davey, Paxman & Co.'s Portable Engine.

49

Messrs Davey, Paxman & Co., Ltd.

Messrs Davey, Paxman & Co., Ltd., of Colchester, have favoured us with an advance sheet of their latest pattern portable engine, Fig. 33.

The cylinder is bolted to a steel seating riveted to the firebox shell, and the crankshaft bearings are likewise carried on steel plate brackets, riveted to the barrel of the boiler; cylinder and plummer-blocks being connected by steel tie-rods with a sliding joint under each bearing to allow for the expansion of the boiler. There are only two small steam joints to be made between engine and boiler, so that there is no difficulty in combining the two, even with unskilled labour, on arrival abroad, if shipped separately, for the sake of a saving in freight. As would naturally be expected, an efficient variable expansion eccentric is fitted, by which the engine can be set to run at any ordinary degree of expansion in either direction.

Paxman's Improved Throttle Valve Governor is shown in section in Fig. 34, but we understand that the Pickering governor may be had in preference if desired, and Fig. 33 shows an engine so fitted.

Messrs Davey, Paxman & Co., Ltd., are justly proud of the performance of their 8 H.P. single-cylinder portable engine at the

4

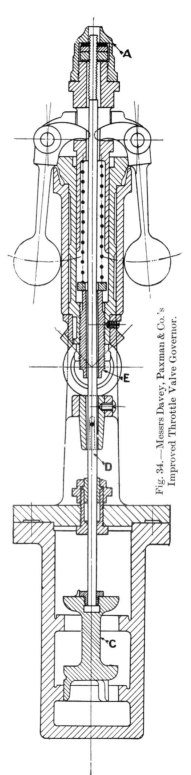

Fig. 34.—Messrs Davey, Paxman & Co.'s Improved Throttle Valve Governor.

Royal Agricultural Society's Newcastle meeting, when they received the prize of £100 for their "Simple" portable engine, the judges' report being as follows:—

"This engine was run at 132 revolutions, and a pressure of 105 lbs., with a brake load of 17 H.P. It ran for 4 hours 23 minutes of actual time with a supply of 193 lbs. of coal, equivalent to a consumption of 2·6 lbs. of coal per horse-power per hour, beating the previous best Cardiff record of 2·79 lbs."

The boilers are made throughout of specially ductile mild steel. The edges of all plates are carefully machined, and all rivet holes are drilled in place, and the rivets closed by hydraulic pressure while the plates are being held together tightly by a secondary hydraulic squeezer.

Messrs Davey, Paxman & Co., Ltd., remind us that they were the first firm in the trade to make their fireboxes of steel; and for many years were the only makers using this material. Seeing that steel has long since ousted the Yorkshire iron, formerly considered the only material (except copper) fit for fireboxes, Messrs Paxman's foresight has been abundantly justified.

Messrs William Foster & Co., Ltd.

Messrs Wm. Foster & Co., Ltd., of Lincoln, are, at the time of writing, "busily engaged on new designs and preparations of patterns for an entirely new type of single-cylinder portable engine, but it is rather premature at present to publish drawings or illustrations of the same." Messrs Foster go on to say that they consider their design will be extremely novel in many directions, and that they would much prefer notice being given of a new type of single-cylinder portable engine, than that any reference should be made to their present type, which type is about to become extinct.

We are looking forward with more than a little interest to the appearance of Messrs Foster's new design, which we are sure will be welcomed by all who are interested, from a purely engineering point of view.

Messrs Richard Garrett & Sons, Ltd.

Messrs Richard Garrett & Sons, Ltd., of Leiston, Suffolk, are amongst the oldest makers of portable engines in this country, and enjoy an enviable reputation for the excellence of their manufactures.

Their single-cylinder portable engine is made in the design shown by Fig. 35, from 6 to 10 N.H.P., and presents many points of interest, being an exception to the usual or Lincolnshire pattern.

This firm are specialists in steel plate flanging. We have already illustrated their corrugated firebox, and now show their flanged steel saddle for the crankshaft plummer-blocks, Fig. 36. Messrs Garrett do not, however, adopt the same idea in connection with the cylinder, the latter being bolted directly to the firebox casing by a heavy and stiff all-round flange, cast on the cylinder body. Neither do they make any use of the stay-rod and sliding plummer-block system, now so largely used; as they are of the opinion that these do not give the necessary stability. To the steel crank saddle are strongly bolted by both vertical and horizontal flanges, Fig. 37, the cast-iron plummer-blocks, which carry two-part brasses, and are inclined towards the cylinder at an angle of about 15 . This angularity has the effect of placing the joint, or parting, of the brasses in such a position, that on the flywheel side both the weight of the wheel and the pull of the driving belt are borne on the solid brass, while on the crank side of the engine the thrust of the crankshaft is taken only on the lower or outward stroke of the piston—assuming the engine to be running "inwards."

We believe that this arrangement, so far as the portable engine is concerned, is peculiar to Messrs Garrett, who are also alone, amongst large makers, in retaining the steam inlet underneath the cylinder, having apparently not adopted the modern plan of a separate external steam bend, the steam joints of which are visible and accessible.

The steam, having entered the cylinder, passes upward and around it, first to the starting valve, and then to a double-seated throttle-valve controlled by a governor of the Pickering type, shown on the cylinder in Fig. 38, and in section in Fig. 39.

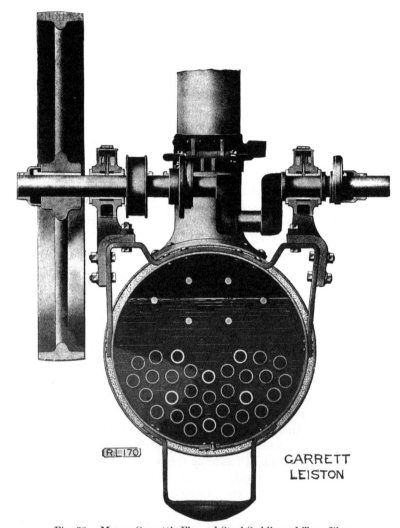

Fig. 36.—Messrs Garrett's Flanged Steel Saddle and Turn-Plate.

STANDARD TYPES

Fig. 38.—Messrs Garrett's Cylinder, with Pickering Governor.

Fig. 37.—Messrs Garrett's Double-Grip Plummer-Block.

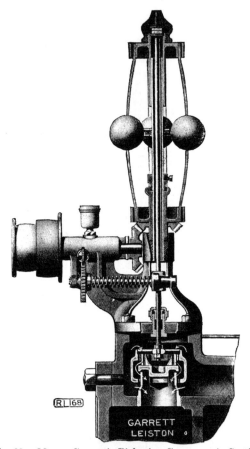

Fig. 39.—Messrs Garrett's Pickering Governor, in Section.

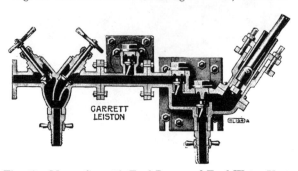

Fig. 40.—Messrs Garrett's Feed-Pump and Feed-Water Heater.

Fig. 35.—Messrs Richard Garrett & Sons' Portable Engine.

Fig. 41.—Messrs Marshall's Portable Engine.

Messrs Garrett's feed pump, Fig. 40, with the suction chamber for drawing in a portion of the exhaust steam, and so raising the temperature of the feed water, is an exceedingly neat and successful application of the injection principle to feed water heating.

Turning now to the wheels and axles, Fig. 36 shows the pressed steel turn plate for the front axle riveted to the boiler barrel as usual, and not, as formerly, to the smoke-box. The hind axles are admirably designed, being socketed into heavy cast-iron flanges, which are secured to the sides of the firebox by studs and nuts.

MESSRS RICHARD HORNSBY & SONS, LTD.

This old-established firm advise us that they are not now manufacturers of portable steam engines. Their portable oil engine, of which they send illustrations, is unfortunately outside the scope of the present work.

MESSRS MARSHALL, SONS & CO., LTD.

Messrs Marshall, Sons & Co., Ltd., whose works at Gainsborough, Lincolnshire, employ four thousand hands, are manufacturers of portable engines upon a very extensive scale. Their engines embody all the essential details pertaining to the highest class of portable engines, and in neatness and finish are unexcelled. Fig. 41 shows their single-cylinder engine as made in the 10-in. to 12-in. cylinder sizes.

The cylinder is mounted on planed steel girders, riveted to the firebox, with an external steam inlet. The flow of steam is carried round the cylinder, and enters the valve chest through the starting valve, which is situated at the top, thus ensuring a dry steam supply.

As usual with all the leading makers, the working barrel of the cylinder is a separate casting of very hard metal, turned and bored, and forced by hydraulic pressure into the body of the cylinder, the space between the two forming the steam jacket. This entirely supersedes the older method of casting the cylinder in one piece.

The crankshaft brackets are each formed by a pair of steel plates riveted together, one flat and the other curved, both being double-riveted to the boiler barrel. On the flywheel side the plummer-block is rigidly bolted to its steel bracket; but on the other side, next to the crank, the plummer-block is capable of sliding in a dove-tailed groove, and is connected by a steel stay-rod to a lug on the side of the cylinder. This forms the provision for allowing the expansion of the boiler without affecting the working parts; and, conversely, for relieving the boiler of the stresses due to the pull-and-thrust of the engine.

Fig. 42.—Messrs Marshall's Pickering Governor.

This arrangement of stay-rod and sliding plummer-block on one side only undoubtedly fulfils the latter purpose perfectly, but inasmuch as one plummer-block partakes of the movement due to the expansion and contraction of the boiler, whilst its companion remains unaffected by these alterations, it is evident that the crankshaft cannot always be at right angles with the axis of the engine. Presumably, however, it is so adjusted as to be correct when the boiler is hot, which would really seem to be all that is necessary. Messrs Marshall's vast experience in the manufacture of portable engines is, moreover, a sufficient guarantee of the practical success of this arrangement.

The governor is again of the Pickering type, and Fig. 42 shows one of these governors with its double-beat throttle-valve just as it is lifted out of the engine.

The feed pump is not attached directly to the boiler plate, but to a steel flanged seating riveted thereto. Attention should be called to the ingenious fore-carriage, which is composed of a single flanged steel plate, in lieu of the wood or angle-iron carriage generally used.

The Marshall boiler is remarkable for the large radius of the rounded corners of the firebox, and is, it is needless to say, an admirable specimen of boiler-work.

A feature in these engines is the mechanical oil pump for lubricating the cylinder, which will be seen in Fig. 41, fixed upon the exhaust pipe, close to the end of the trunk guide casting.

MESSRS RANSOMES, SIMS, & JEFFERIES, LTD.

Messrs Ransomes, Sims, & Jefferies, Ltd., of the Orwell Works, Ipswich, are, as our historical section will show, among the oldest makers of portable engines in the country, having received the Royal Agricultural Society's first prize at the Bristol meeting in the year 1842.

Their present-day engine is fully up to date, however, as shown in Fig. 43.

Messrs Ransomes employ the steel brackets for carrying the crankshaft, with stay-rod and sliding plummer-block on the crank side only, the cylinder having the modern external steam inlet and jacket feed inlet; but an adherence to the older design is traceable in the formation of the cylinder, which is curved to fit the boiler. The four slide-bars for carrying the crosshead guide-blocks, though largely abandoned by other eminent makers nowadays, have still much in their favour. The bolted-on connecting rod straps—a system used almost universally for locomotive engines in the interests of safety—are a distinct improvement on the ordinary pattern. The firm's own high-speed governor, or the Pickering type, may be fitted at pleasure, and that very convenient attachment, a ladder, is fitted upon all engines of 6 N.H.P. and upwards. A ball-and-socket arrangement permits the fore-carriage to tilt as required on uneven ground. We may conclude by a word of praise for the large hand-hole shown on the right-hand side of the firebox, with a good strengthening ring riveted round it.

MESSRS ROBEY & CO., LTD.

The portable engine manufactured by Messrs Robey & Co., Ltd., of Lincoln, may fairly claim, on its merits, to stand in the very front rank. The cylinder is bolted to a planed steel bracket, riveted to the firebox casing, as are also the crankshaft brackets; not a single bolt being employed to attach any of the working parts to the boiler, the intermediate support for carrying the piston rod guide-bars being formed by a cast-iron motion plate spanning the two stay-rods. Thus, no part of the engine is affected by the varying length of the boiler as it is heated and cooled.

58 THE PORTABLE STEAM ENGINE

Fig. 43.—Messrs Ransomes' Portable Engine.

Fig. 44.—Messrs Robey's Portable Engine.

The external steam bend through which the main supply is taken to the cylinder not only contains the starting valve, but, being extended upwards, forms a seating for the Ramsbottom safety-valve—a particularly neat and compact arrangement. Even the pressure gauge is attached to this casting, so that the number of steam joints in connection with the boiler shell is reduced to the absolute minimum. Thus the feed pump is attached to the steel crankshaft bracket, the feed water inlet itself being the only fixing actually connected to the boiler.

We have already illustrated Messrs Robey's method of staying the firebox crown (Fig. 24), which relieves the firebox from the crushing stresses tending to distort the front and back plates, and allows of free access for cleaning purposes to every part of the crown plate—a very important matter in cases where incrustation is present.

This is a good engine, strikingly simple and direct in the application of all its parts, and well calculated to stand not only the rough usage generally accorded to this class of engine, but also to give effect to the high boiler pressures now employed.

Messrs Robey also furnish us for comparison Fig. 45, which shows their former type, now discarded for even the smallest sizes in favour of the modern design of Fig. 44.

Messrs Ruston, Proctor & Co., Ltd.

Messrs Ruston, Proctor & Co., Ltd., of Lincoln, send us a photograph of their latest type of single-cylinder portable engine, Fig. 46, into which a very remarkable development is introduced. The idea itself is not a new one, but it is safe to say that its revival in the form adopted by Messrs Ruston is only rendered possible by the use of modern machine tools of exceptional size and power.

To the top of the firebox casing is riveted a large square cast-steel seating of size equal to the flat bottom of the cylinder. The included portion of the firebox top has a large oval hole cut through the plate; and upon this seating the cylinder is bolted by a strong and wide flange, passing all round the latter; the combination forming, to all intents and purposes, a square steam dome having the cylinder formed within it; the boiler steam,

Fig. 45.—Messrs Robey's Former Method of Construction (for comparison).

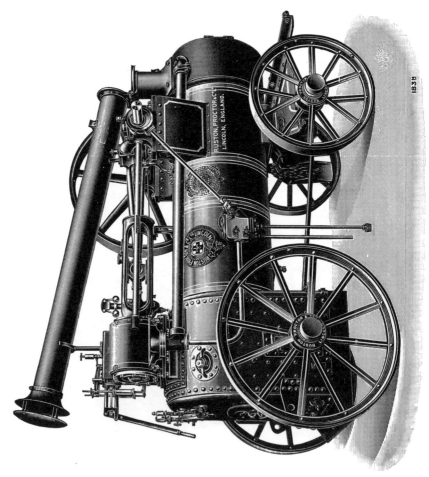

Fig. 46.—Messrs Ruston's Portable Engine.

through the oval hole already mentioned, being in free communication with the steam jacket right round the liner which forms the cylinder barrel, Fig. 47.

Of the strength and solidity of this arrangement there can be no two opinions; that the huge steam joint underneath the cylinder will give no trouble is practically guaranteed by the fact that the entire boiler, after the seating and brackets are riveted upon it, is put into a planing machine, and the flat surfaces for the reception of the cylinder and plummer-blocks machined to a true and level plane.

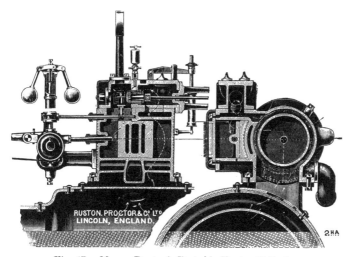

Fig. 47.—Messrs Ruston's Portable Engine Cylinder.

Our notice of this engine would not be complete if we omitted to mention their famous steam-heated expanding stay—a tubular tie-rod, connecting cylinder and one bracket, in free communication with the boiler at both ends, with the intention of keeping pace with the varying length of the latter. The small pipes connecting the interior of the tubular stay with the boiler are visible in the illustration, Fig. 46.

It would seem to be a defect in this arrangement that, until steam is actually formed, no more expansion of the stay takes place than would occur in the ordinary course if the stay were solid, and heated merely by conduction. The temperature of saturated steam of 100 lbs. pressure above the atmosphere is 328° Fahr.; and assuming that during the process of steam raising the stay has heated by conduction to 100° Fahr.—a liberal estimate—while the boiler has reached 212° Fahr., we have, at the boiling point, a difference of 112° Fahr., the stay being operative only during the remaining 116° Fahr. This may, or may not, be a practical objection; at any rate, theoretically speaking, it only meets the difficulty half way.

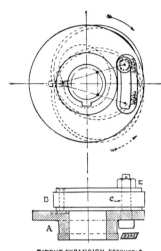

PATENT EXPANSION ECCENTRIC.
Fig. 48.—Messrs Ruston's Patent Expansion Eccentric.

It should be noted that the stay is only applied to one of the crankshaft bearings — that next the crank—and that, of course, no provision for sliding the plummer-blocks is necessary. Messrs Ruston's adjustable eccentric for varying the point of cut-off is shown in Fig. 48.

Messrs E. R. & F. Turner, Ltd.

Messrs E. R. & F. Turner, Ltd., of St Peter's and Grey Friars Works, Ipswich, are a firm of very old standing, who, as far back as 1872, competed at the Cardiff trials of the Royal Agricultural Society with very marked success as regards economy of fuel, 2·9 lbs. of coal per indicated horse-power per hour being their official record.

For many years this firm, almost alone amongst British manufacturers, recognised the importance of a direct connection between crankshaft and cylinder, and what is more, regularly carried it out in their engines. Not, it is true, exactly in the modern form, but in a manner so ingenious as to deserve placing upon record here.

The crankshaft plummer-blocks were each mounted upon a narrow steel plate riveted edgewise to the boiler (i.e., so as to present only a thin edge to the observer standing exactly in line

STANDARD TYPES

Fig. 49.—Messrs Turner's Portable Engine.

with the axis of the crankshaft), thus forming an elastic—as distinguished from a sliding—crank bearing. From these extended to the cylinder upon each side a flat bar, forming a rigid stay, and absorbing all the pull-and-thrust stresses of the piston.

Really, except upon the ground of appearance—the crankshaft supports being almost invisible on a strictly broadside view at some little distance away—it is difficult to imagine why this simple and effective arrangement should have given place to the more modern, but far less original design shown in Fig. 49, which represents Messrs Turner's present-day engine.

However this may be, Messrs Turner have formed into line with most of their competitors, and their engine now exhibits no peculiarities. Even the Turner-Hartnell governor has gone by the board, leaving us, it is true, with a first-rate modern steel-bracketed detachable engine on a flush-topped locomotive boiler, but—well, an ancient landmark has disappeared, and we are half inclined to regret it.

CHAPTER IV

THE COMPOUND PORTABLE ENGINE

HITHERTO we have been considering what may be called the rank and file—the ordinary stock portable engine of commerce. We now enter upon another section of our subject, where the scope for individuality in design is much greater—the high and low pressure, or compound, portable engine.

Power upon wheels is such a desideratum that it is not surprising to find the portable engine, once an exclusively agricultural machine, regularly adopted for industrial purposes, with the consequence that in dimensions, no less than in mechanical efficiency, it has developed "out of all knowledge." Engines of 200 effective H.P., though not exactly common, are to be had, as a regularly catalogued product, of some of our leading firms.

The process of evolution which has produced the type we are now dealing with has undoubtedly been through the double-cylinder class. Portable engines with two equal high-pressure cylinders have been in use for very many years—in fact, a double-cylinder engine upon wheels is recorded among the exhibits entered at the Royal Agricultural Society's meeting in the year 1847. The use of two-cylindered engines was expressly discouraged by the Society on the ground of complexity—and rightly so, for the small powers then in use. Large sizes, however, being speedily called for, the double-cylinder engine came to the front again; and, until the year 1879, was all but universal for engines of anything above 12 or 14 N.H.P.

In that year, Messrs Garrett, of Leiston, introduced the compound principle for portable steam engines; and thus gained the credit of initiating a remarkable improvement. Their experiment in this direction proved successful, and the compound portable engine soon became established as a standard production by most of the leading makers.

There is a good deal of misapprehension amongst users of steam power—and others—as to the relative values of compound and simple single or double cylinder engines; and occasionally claims so extravagant as to defeat their own object are put forth by those who ought to know better.

We will here do our best to give a plain answer to a plain question. What is there in a compound engine—speaking now of portables and their allied types—to make up for the increase in first cost and maintenance, for the additional weight and complication, and for the extra risk from higher speeds and higher steam pressures? There is only one answer to such a question. Increased economy—a better result per pound of fuel used.

In determining, therefore, whether, in given circumstances, it is or it is not worth while to purchase a compound engine in preference to a simple, whether single or double cylinder, practically the only consideration will be the cost of the fuel; the difference in the first cost of the two types being in comparison almost negligible.

This seems, at first sight, rather a strong assertion, but let us examine it further. First, as to fuel consumed. Starting with the purely practical side of the question, and referring the reader to any of the numerous and excellent text-books on the subject for the theoretical proof, we may define the consumption of fuel, in the two cases of equally well-designed simple and compound portable engines, as represented by the figures 35 and 25 respectively—a saving of 10 in every 35, or 28·6 per cent. This is our own experience.

Mr Walter Hutton in his carefully collected "Tables of Economic Performances" of various kinds of steam engines, gives 27·8 per cent. Finally, the average saving, as collated from the various figures supplied by the makers of the engines we are illustrating, amounts to 29 per cent. Now, if we take the mean of all these, we get a figure of $28\frac{1}{2}$ per cent., as the actual saving.

Basing our calculation on 100 B.H.P., ten hours daily for 300 working days per annum, we have as the cost for fuel of a well-designed, non-compound engine and boiler, using 4·375 lbs.

of coal per brake horse-power per hour, say, 586 tons per annum at 12s., or £351. 12s. A saving of 28½ per cent. upon this amounts to the respectable sum of almost exactly £100 a year. In other words, with coal at 12s. per ton one can save £1 per annum upon each brake horse-power by compounding.

Water also generally costs something; the amount of water saved is nearly proportional to that of the fuel; but we may set against this the additional cost for oil and consumable stores, and the greater wear and tear from the higher speed and pressure employed. It is impossible to pin these latter items down to any definite figures, as the cost of water may vary from next to nothing where it is raised from a well on the premises, or drawn from a river, up to a considerable sum where supplied from the town mains. In the former case, where a plentiful supply exists, a condenser may be added, which will still further reduce the coal bill from 15 to 20 per cent.

In view of such an economy in fuel as is here presented, the statement that the extra first cost of the higher class engine is almost negligible, is, we think, fully sustained. There are, of course, circumstances in which no one would think of employing a compound engine. In the case of a saw-mill, for instance, where the cost of fuel may be considered as little or nothing; or in small sizes, where the saving effected, though relatively proportionate, is actually small.

Roughly speaking, the difference in cost, as between a double-cylinder, high-pressure, portable engine and a compound—or high and low pressure engine—of equal effective power, averages 26 per cent., which cannot be regarded as an extravagant addition to the first cost, considering the advantages it secures.

Now we may proceed to consider very briefly the various makes of compound portable engines which British manufacturers have to offer. In making comparisons, the nominal horse-power should be disregarded entirely, and only the effective or brake horse-power considered. Again, most makers give two effective horse-powers, the lower as representing the normal or daily load for which their engine is suitable; the larger figure denoting the occasional or emergency load, in other words, the maximum power available for short periods.

THE PORTABLE STEAM ENGINE

Messrs Brown & May, Ltd.

Proceeding alphabetically as before, we illustrate in Fig. 50 Messrs Brown and May's (Ltd.) compound portable engine. Their tabulated dimensions and powers are given in the annexed table:—

Compound Portables.

Nominal Horse-Power.	8	10	12	16	20	25
Brake horse-power—						
Working load - - - - -	24	30	36	48	60	75
Maximum load - - - - -	28	35	42	56	70	88
Working steam pressure, in lbs. per square inch - - - - - - -	140	140	140	140	140	140
Diameter of high-pressure cylinder, in inches - - - - - - -	5½	6¼	7	8	9	10
Diameter of low-pressure cylinder, in inches - - - - - - -	9¼	10½	12	13	14¼	16
Length of stroke, in inches - - -	12	12	12	14	14	18
Speed, in revolutions per minute - -	180	180	180	155	155	120
Diameter of flywheel, in inches - -	54	54	54	66	66	72
Width of turned face on rim, in inches -	8	8	9	12	12	14
Weight on wheels unpacked, in cwt. (about)	115	126	136	185	220	300

Lack of space alone prevents our giving a detailed description of this engine, which the makers state is the most economical engine obtainable; and with regret we must content ourselves with referring intending purchasers to the firm at Devizes, who will doubtless be able to make good their assertion.

Messrs Clayton & Shuttleworth, Ltd.

Messrs Clayton & Shuttleworth, Ltd., manufacture their compound portable engines in sizes from 10 to 30 N.H.P., in accordance with the annexed illustration, Fig. 51, from which it will be seen that the flush-topped boiler (working at 140 lbs. pressure) is strongly stayed by five longitudinal bolts, and fitted with a steel manhole cover. The cylinders are bolted to a planed saddle secured to the arch-plate of the boiler, and the only steam joint is the admission elbow to the cylinder, which can be removed without disturbing any other part. Both the high and low

COMPOUND ENGINES

Fig. 50.—Messrs Brown & May's Compound Portable Engine

Fig. 51.—Messrs Clayton & Shuttleworth's Compound Portable Engine.

pressure steam chests are outside (*i.e.*, not between) their respective cylinders, and are consequently perfectly accessible for refacing the valve surfaces when required, an advantage by no means to be overlooked. On the other hand the additional width between the two crankshaft bearings arising from this arrangement seems to call for a central bearing, for which, we observe, there is just room between the two cranks. In the larger sizes, we are glad to see, this central bearing is provided, all three being mounted on a saddle-bracket bolted to the boiler barrel.

Messrs Clayton & Shuttleworth inform us that they are prepared to supply either automatic expansion gear or the well-known Pickering governor, acting upon a throttle-valve, for these engines.

Messrs Davey, Paxman & Co., Ltd.

Messrs Davey, Paxman & Co., Ltd., of Colchester, build their compound portable engines upon somewhat different lines; carrying the detachable principle still further than any of their competitors (Fig. 52).

Instead of bolting their cylinder and the three bearings of the crankshaft directly to flanged steel seatings riveted to the boiler, the complete engine is fully erected upon a steel girder-frame, which, in its turn, is secured to the boiler brackets. Of course, in this case, the steel tie-rods are omitted; the stresses being taken by the girder-frame. Messrs Paxman's compound portable engine has a splendid record, as we shall presently see; but we cannot help saying that, structurally speaking, we think the straight line principle embodied in the simple tie-rod, carried on a level with the axis of the engine, preferable to the heavier and more elaborate girder-frame, several inches below the centre line.

However, as in neither case is there likely to be any yielding or longitudinal weakness, we may treat this as a matter of individual opinion, and pass on to other matters. The crankshaft is not machined from a solid steel forging, as is usual in this class of engine, but is bent from a round steel bar. Probably the bent crankshaft is strong enough, but we must confess to a liking for the bright steel "throws" of the slab pattern of crankshaft.

The automatic expansion gear, which is part of the regular equipment of these engines, is on the well-known Paxman system,

Fig. 52.—Messrs Davey, Paxman & Co.'s Compound Portable Engine.

COMPOUND ENGINES 73

two eccentrics being employed to actuate the expansion valve; which, by the way, does not slide directly on the back of the main valve, but upon a cast-iron plate, having the steam ports cut through it, sandwiched between the two valves. The rods from these two eccentrics are coupled to a link exactly as in the Stephenson reversing gear; and the governor raises and lowers

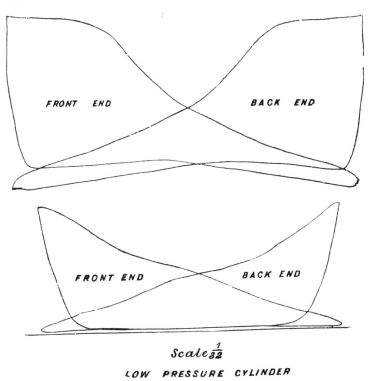

$Scale \frac{1}{32}$

LOW PRESSURE CYLINDER

Fig. 53.—Diagrams from Messrs Davey, Paxman & Co.'s Valve Gear.

this link as the load varies, the valve rod partaking more or less of the motion of either eccentric rod according to the height of the governor sleeve. The position and travel of the eccentric actuating the bottom of the link are such as to give a very early cut-off; the movement of the top of the link, derived from the other eccentric, giving the maximum cut-off point. Diagrams taken from an engine fitted with this gear are shown in Fig. 53, and the

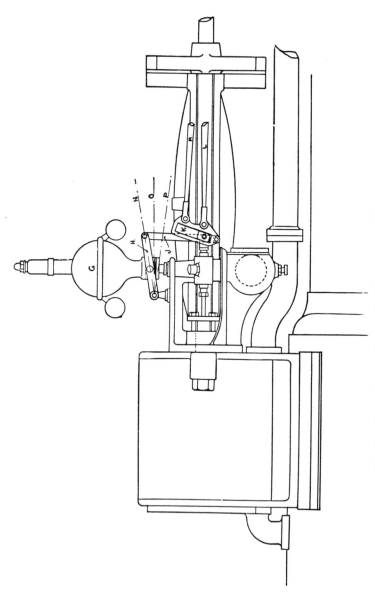

Fig. 54.—Messrs Davey, Paxman & Co.'s Automatic Expansion Gear.

gear itself in Fig. 54. The governor is of the dead-weight type, with friction rollers interposed between the arms and the weight, thus greatly increasing the sensitiveness of its movement.

The principal dimensions, as tabulated by the makers, are as follows: the ratio between the cylinders averaging 2·6 to 1. The working pressure is 140 lbs.

Nominal Horse-Power.	Brake Horse-Power.		Diameter of Cylinders.		Stroke Both Cylinders.	Flywheel.		Revolutions per Minute.
	Economical Load.	Maximum Load.	High Pressure.	Low Pressure.		Diameter.	Face.	
	H.P.	H.P.	In.	In.	In.	Ft. In.	In.	
8	19	23	5½	9	14	5 0	7	155
10	26	30	6½	10½	14	5 0	8	155
12	30	35	7	11¼	14	5 0	9	155
16	40	48	8	13	14	5 6	10	155
20	50	60	9	14½	16	6 0	11	135
25	65	78	10	16	18	7 0	12	120
30	75	90	11	17½	18	7 0	14	120

The Royal Agricultural Society's prize of £200 was awarded to this firm in July 1887, for their 8 H.P. (nominal) compound portable engine, on its recorded consumption of 1·85 lbs. of coal per brake horse-power per hour; beating their own single-cylinder engine of the same nominal horse-power, which consumed 2·6 lbs., almost exactly in the proportion of 25 to 35; thus confirming the comparative figures given in the earlier part of the present section, from independent sources, as the measure of the relative economy of the compound and high-pressure types.

MESSRS WILLIAM FOSTER & CO., LTD.

Messrs William Foster & Co., Ltd., of Lincoln, construct their compound portable engines, Fig. 55, for a working pressure of 200 lbs. per square inch; and as will be seen from the table below, their speeds also are very much higher than is usual in this class of engine.

Fig. 55.—Messrs Wm. Foster & Co.'s Compound Portable Engine.

COMPOUND ENGINES 77

TABLE OF DIMENSIONS OF MESSRS WM. FOSTER & CO.'S
COMPOUND PORTABLE ENGINE.

Size or Mark.	ENGLISH DIMENSIONS.					
	Cylinders.			Flywheel.		
	High Pressure.	Low Pressure.	Stroke of Piston.	Diameter.	Width.	Revs. per Minute.
	In.	In.	In.	Ft. In.	In.	
C.P. 6 - -	$4\frac{1}{2}$	$6\frac{3}{4}$	9	2 6	4	250
C.P. 8 - -	$4\frac{3}{4}$	$7\frac{1}{2}$	9	3 0	5	250
C.P. 10 - -	$5\frac{1}{2}$	$8\frac{1}{2}$	10	4 0	6	225
C.P. 12 - -	6	9	10	4 0	6	225
C.P. 15 - -	6	$9\frac{1}{2}$	12	4 6	7	200
C.P. 18 - -	$6\frac{3}{4}$	$10\frac{3}{4}$	12	4 6	7	200
C.P. 20 - -	$7\frac{1}{4}$	$11\frac{1}{2}$	12	4 6	$8\frac{1}{2}$	200
C.P. 25 - -	$7\frac{3}{4}$	12	12	4 6	$9\frac{1}{2}$	200

For this reason they claim to produce a given power from an engine of smaller weight and dimensions than is possible with the general type of compound engine. This advantage in weight and dimensions is, of course, a considerable item in hauling the engine over long stretches of rough country; also the freight for export is correspondingly reduced.

The cylinders are mounted on a planed steel seating which is riveted to the boiler; and the crank brackets are also of steel, fitted with brasses

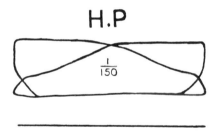
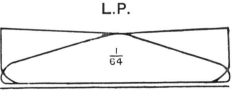

Fig. 56.—Diagrams taken from Messrs Wm. Foster & Co.'s Compound Portable Engine, with Pickering Governor.

adjustable in four directions. No tie-rods are used. Messrs Foster have also abolished the automatic expansion gear on their compound engines in favour of the Pickering governor and throttle-valve: relying on their tubular feed-water heater for making up the economical effect. That they are working on right lines in this direction cannot be doubted; the diagrams shown in Fig. 56 having been taken from an engine so fitted, with cylinders $\frac{6+9\frac{1}{2}}{12}$, at 200 revolutions per minute, developing 40 B.H.P. This engine, on a six hours' test recently, gave: steam 23 lbs. per brake horse-power, coal 2·9 lbs., water evaporated per lb. of coal 8 lbs., at 200 lbs. working pressure, which is an excellent result, and goes far to prove the modern tendency to simplify the engine, by substituting a plain governor for the complexities of the automatic cut-off system, to be correct.

Messrs Richard Garrett & Sons, Ltd.

Messrs Richard Garrett & Sons, Ltd., of Leiston, make their compound portable engines up to 42/48 B.H.P., as Fig. 57; these figures representing respectively the economical and the maximum brake horse-powers. Above these powers, and up to 80/100 B.H.P., they are constructed as Fig. 58. The cylinders in both cases are bolted to flat flanged seatings; but curiously, these are in cast iron (as is also the crank saddle carrying the three bearings, Fig. 59) in the larger sized engines, while in the smaller they are of flanged steel plate riveted to the boiler. In Fig. 58, the boiler is "flush-topped"; but otherwise, so far as we can see, the designs are similar right through. Messrs Garrett do not now fit automatic expansion gear to their engines, except at an extra cost when ordered specially, preferring to make use of the Pickering throttle-valve governor in all ordinary cases. The Rider expansion governor, Fig. 60 used by Messrs Garrett when automatic cut-off gear is specified, is interesting as being an English invention used far more largely on the Continent than in this country, where, we believe, Messrs Garrett are the only prominent firm to adopt it.

In this gear the outer or expansion slide-valve remains at constant travel, but is adjusted by the governor for varying points

Fig. 57.—Messrs Garrett's Compound Portable Engine. Sizes up to 48 max. B.H.P.

Fig. 58.—Messrs Garrett's Compound Portable Engine. Sizes from 55 to 100 max. B.H.P.

COMPOUND ENGINES 79

of cut-off through a rack and pinion arrangement, which rotates the valve spindle, thus moving the outer valve up or down on the inner or main valve, so as to give a nearly uniform speed whatever the conditions of load or pressure.

Messrs Garrett were the first to apply the compound principle to portable engines, and were awarded special prizes of honour

Fig. 59.—Messrs Garrett's Cast-iron Saddle for their larger-sized Engines.

at the R.A.S.E. show at Carlisle in 1879, and at the Inventions Exhibition in London in 1881, for the first compound portable engine made and exhibited. To this day Messrs Garrett's engines differ radically in points of detail from those of all other makers, as will be apparent to the most casual observer. To give a single instance the tail rod, or prolongation of the piston rod through the

Fig 60.—The Rider Automatic Expansion Gear as Used by Messrs Garrett.

Fig. 61.—Messrs Marshall's Compound Portable Engine.

COMPOUND ENGINES 81

back cylinder cover—a feature in all engines of this firm of 10 N.H.P. and upwards—is practically unknown outside the Leiston works.

In conclusion we must congratulate Messrs Richard Garrett & Sons, Ltd., on their handsomely turned-out catalogues, which are so endowed with a wealth of illustration and description as to make it matter for regret that we cannot devote more space to their productions.

MESSRS MARSHALL, SONS & CO., LTD.

Messrs Marshall, Sons & Co., Ltd., of Gainsborough, build their compound portable engines on the lines shown in Fig. 61; that is to say, the cylinders and crank bearings (of which there are three) are bolted to steel brackets riveted to the boiler, and connected by

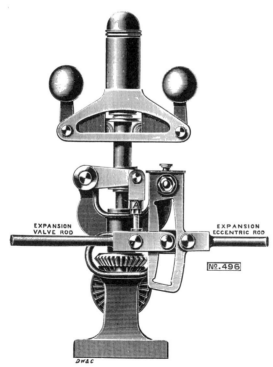

Fig. 62.—The Hartnell Patent Governor and Automatic Expansion Gear. Messrs Marshall, Sons & Co., Ltd.

tie-rods; the whole being mounted on a flush-topped boiler, designed for a working pressure of 150 lbs.

The Pickering governor is now a standard fitment upon these engines; but where automatic expansion gear is specified, Messrs Marshall apply the Hartnell gear, as illustrated in Fig. 62, which has had a long and successful career, and is very compact and complete. Steam is supplied to the cylinder, through an external copper connection, from a steel seating riveted to the boiler. There is a drain pipe from the cylinder jackets, which is carried to the smoke-box end of the boiler to facilitate the circulation of the jacket steam—a precaution, praiseworthy in itself, but not, one would imagine, of any great service actually. Still it is an index of the thoroughness, in matters small as well as great, for which the Britannia Works is celebrated. There is also a lavishness, almost amounting to luxury, in small details which is very pleasing to the engineer's eye.

Messrs Ransomes, Sims & Jefferies, Ltd.

Messrs Ransomes, Sims & Jefferies' compound portable engine is shown in Fig. 63.

From 8 to 20 N.H.P. the engines are constructed to this design; the 25 and 30 N.H.P. engines being built on a girder-frame which is fixed on brackets on the boiler so that at any time the engine may be detached, and set to work as an independent engine. Automatic expansion gear, of the type shown in Fig. 64, is supplied in their standard practice; but the single slide-valve and Pickering governor, if preferred, are fitted at a reduction of price. The crank brackets (two) are of wrought iron, riveted to the boiler, and connected (in the smaller sizes) to the cylinders by steel tie-rods. They are provided with gun-metal bearings having a vertical and lateral adjustment. The crank itself is bent out of a round bar of steel. The steam chests are outside their respective cylinders, thus ensuring easy access; and the steam inlet is external, so that there is no steam joint under the cylinders.

Messrs Ransomes, Sims & Jefferies' name has been identified with the portable steam engine since its earliest days, as a reference to our historical pages will show; and their engines have always kept pace with the times.

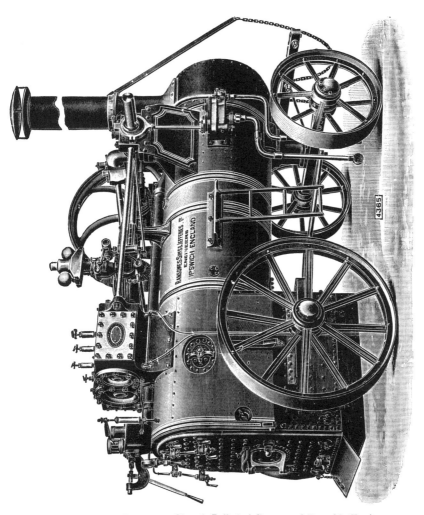

Fig. 63.—Messrs Ransomes, Sims & Jefferies' Compound Portable Engine.

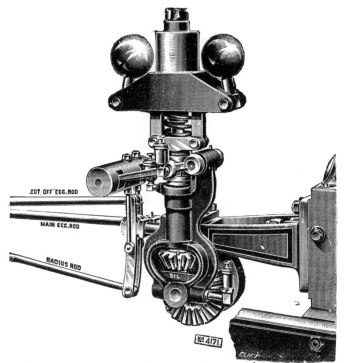

Fig. 64.—Messrs Ransomes' Automatic Governor Expansion Gear.

The principal dimensions of Messrs Ransomes' compound portable engines are as under:—

Nominal H.P.	CYLINDERS.			FLYWHEEL.			
	Diameter, High Pressure.	Diameter, Low Pressure.	Stroke of both Cylinders.	Diameter.		Face.	Revolutions per Minute.
	In.	In.	In.	Ft.	In.	In.	
8	5¾	9	12	5	0	7	180
10	6½	10	12	5	0	8	180
12	7	11	14	5	6	8	155
16	8	12¾	16	5	6	10	135
20	9	14	16	6	0½	10	135
25	10	16	18	6	1½	14	120
30	11	17½	18	6	1½	15	120

COMPOUND ENGINES

MESSRS ROBEY & CO., LTD.

Messrs Robey & Co., Ltd., of Lincoln, are specialists in the compound portable engine, and carry out the detachable principle very thoroughly.

Fig. 65 shows their standard engine, fitted with automatic expansion gear, and crankshaft of forged steel with counter-balance weights.

In response to a demand for a compound engine which should be as simple and easy to manage as an ordinary double-cylinder engine, they have produced the class shown in Fig. 66, fitted with the Pickering governor, and having the crankshaft bent from a solid steel bar; the engines being otherwise identical in construction throughout. Messrs Robey have kindly supplied us with very clear, and almost self-explanatory, drawings of their automatic expansion gear, as applied to a compound portable engine, consisting of a belt-driven, high-speed governor, spring-loaded, capable of varying the position of a die in the swinging link, and thus modifying the travel of the expansion valve, as clearly seen in Fig. 67.

The construction of the double-ported main and expansion slide-valves is also very evident in Fig. 68. This is a very simple and well-arranged expansion gear, and the pair of high and low pressure cards taken from one of their engines thus fitted, shown in Fig. 69, are convincing testimony of its efficiency in action. Messrs Robey's dimension table is annexed.

No.	Effective Horse-Power.		Cylinder.			Flywheel.		
	Economical	Maximum.	High Pressure Diam.	Low Pressure Diam.	Stroke.	Diameter.	Width.	Revs. per Minute.
			In.	In.	In.	Ft. In.	In.	
8	19	26	$5\frac{1}{2}$	$9\frac{1}{2}$	12	4 10	9	200
10	24	32	6	$10\frac{1}{4}$	14	4 10	9	172
12	28	38	$6\frac{1}{2}$	$11\frac{1}{4}$	14	5 0	9	172
14	33	44	$7\frac{1}{4}$	$12\frac{1}{2}$	16	5 6	11	150
16	38	50	$7\frac{1}{2}$	13	16	5 6	11	150
20	48	64	$8\frac{1}{2}$	$14\frac{3}{4}$	16	6 0	12	150
25	60	80	$9\frac{1}{2}$	$16\frac{1}{2}$	18	6 6	13	133
30	72	96	$10\frac{1}{4}$	18	18	7 0	16	133

86 THE PORTABLE STEAM ENGINE

Fig. 65.—Messrs Robey's Compound Portable Engine, with Automatic Expansion Gear and Balance Crank.

COMPOUND ENGINES

Fig. 66.—Messrs Robey's Compound Portable Engine, with Pickering Governor.

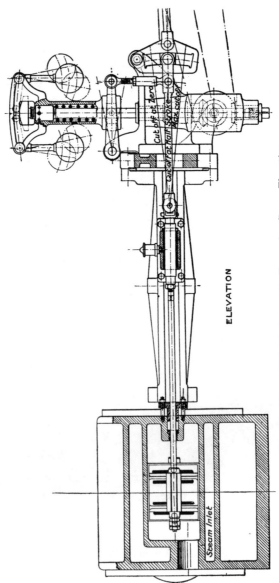

Fig. 67.—Messrs Robey's Automatic Expansion Gear: Elevation.

COMPOUND ENGINES

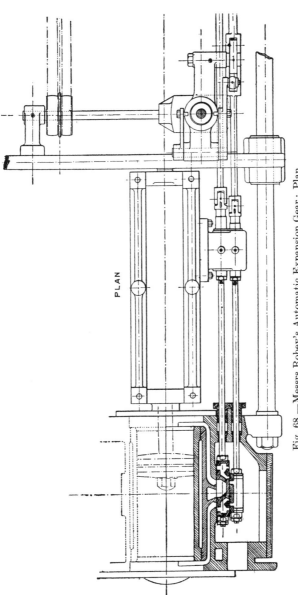

Fig. 68.—Messrs Robey's Automatic Expansion Gear: Plan.

The cylinder ratio is 3 to 1, and the working pressure is 150 lbs. It will be seen that Messrs Robey & Co. go in for a higher rotational speed and a larger cylinder ratio than most of their competitors adopt, and that their margin or reserve of

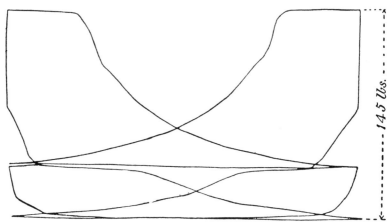

Fig. 69.—Diagrams from Compound Portable Engine. Messrs Robey & Co., Ltd.

effective power is also greater than that tabulated by most other manufacturers of the corresponding class of engine. A 200 H.P. compound portable engine by Messrs Robey & Co., Ltd., is shown in Fig. 70, which is believed to be the largest portable engine ever constructed.

Messrs Ruston, Proctor & Co., Ltd.

Messrs Ruston, Proctor & Co., Ltd., of Lincoln, construct their compound portable engines from 8 to 30 N.H.P. on the detachable principle already fully discussed in connection with their single-cylinder engines. Up to, and including, the 20 N.H.P. size they are as Fig. 71; the 25 and 30 N.H.P. sizes being similar in design, but with outside steam connections to the cylinders. All of these can be fitted either with the Pickering governor acting upon a throttle-valve, or, if desired, with automatic expansion gear. If the latter, the two larger sizes would have link expansion gear; but the Rider type of automatic gear shown in Fig. 72 would be

Fig. 70.—200 I.H.P. Compound Portable Engine.   Messrs Robey & Co., Ltd.

Fig. 78.—Messrs Garrett's Special High-Pressure Single-Cylinder Portable Engine.

COMPOUND ENGINES

Fig. 71.—Messrs Ruston, Proctor & Co.'s Compound Portable Engine. Sizes 8 to 20 N.H.P. (Sizes 25 and 30 are similar, but with Outside Steam Connections to Cylinder).

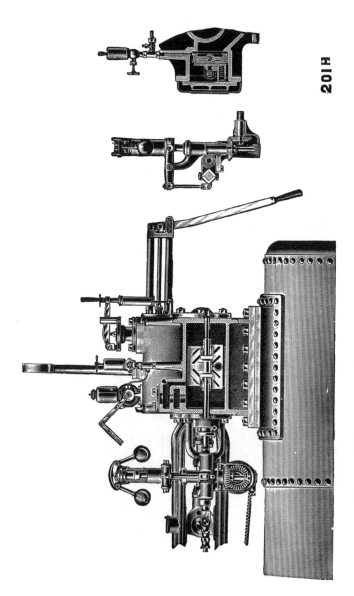

Fig. 72.—Rider Automatic Expansion Gear. Messrs Ruston Proctor & Co., Ltd.

COMPOUND ENGINES 93

used in the sizes up to 20 N.H.P., which may be thus shortly described. The gear consists of an expansion valve with diagonal ports sliding across similar ports on the back of the main slide-valve, each being driven by a separate eccentric. When the governor rises or falls, through an alteration in the load or the steam pressure, a twisting movement is imparted to the valve-spindle which causes a corresponding vertical movement of the expansion valve, and thus causes an earlier or a later cut-off in the cylinder. Diagrams taken from an engine thus fitted are shown in Fig. 73, the dotted lines in each showing the effect of a condenser in economising steam consumption. Messrs Ruston's principal dimensions are as under:—

TABLE OF PRINCIPAL DIMENSIONS OF COMPOUND PORTABLE ENGINES. MESSRS RUSTON, PROCTOR & CO., LTD.

Nominal H.P.	Dimensions.						Effective H.P.	
	Cylinders.			Flywheel.				
	Diameter, High Pressure	Diameter, Low Pressure.	Stroke of both Cylinders.	Diameter.		Face.	Revs. per Minute.	Economical Load.	Full Load.
	In.	In.	In.	Ft.	In.	In.			
8	5¾	9	12	5	0	7	170	18	21
10	6½	10½	12	5	1	7	170	24	29
12	7	11	14	5	6	8	155	30	35
16	8	12¾	16	5	9	9	135	41	48
20	9	14	16	6	0	10	135	50	60
25	10	16	18	6	3	12	120	68	78
30	11	17	18	6	6	14	120	77	90

Referring back to Fig. 71, it will be seen that there are three bearings to the forged steel crankshaft, that tubular or trunk guides are employed, and that the patent steam-heated expanding stay-rod is used between the crank brackets and the cylinder. The boilers of all the sizes are of the flush-top form, and the lagging, or covering of the boiler is, as is usual, carried over the firebox. Both steam chests are outside their cylinders, and the valves are, therefore, easily accessible.

Messrs Ruston, Proctor & Co., Ltd., publish in their catalogue

94 THE PORTABLE STEAM ENGINE

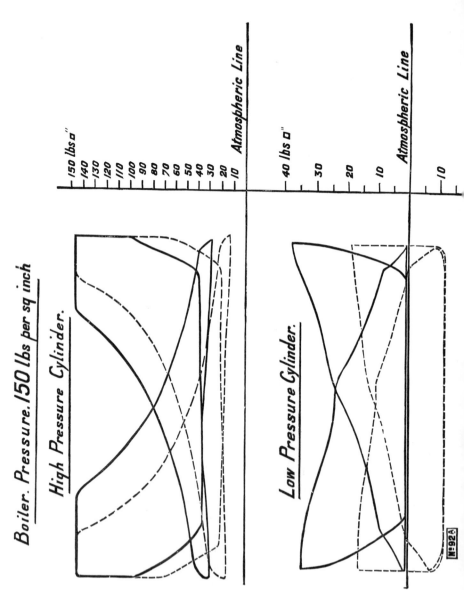

Fig. 73.—Indicator Diagrams from Compound Portable Engines with Automatic Expansion Gear.
Messrs Ruston, Proctor & Co., Ltd.

a *facsimile* letter from a foreign customer of theirs, dated 31/12/07, to the effect that their portable steam engine, No. 714, delivered in 1865—forty-two years earlier—is still at work, "and can last for many other years, the said engine was always in continual good work." This is a convincing testimony of the lasting properties of Messrs Ruston's engines.

In the various examples of compound portable engines we have been considering, it would be difficult, when all are so uniformly good, to award the palm to any one maker, though it would be, comparatively, an easy matter to suggest the design of an engine which should in itself embody all the best points of the collection.

A triple-expansion portable engine has yet, we believe, to be produced—probably the first cost and the weight of such an engine would prevent its adoption to such an extent as to warrant the cost of production—but we may see, from the examples of special types of portable engine still to be examined, that additions to, and modifications of, the compound type, are now being put forward by makers of repute, which will tend still further to the great desideratum of economical steam production and consumption. It is well that we are not standing still, for our foreign competitors are straining every nerve in the race for the world's supply of portable steam engines.

CHAPTER V

SOME SPECIAL TYPES OF PORTABLE ENGINE

ON looking over the illustrations and descriptive matter kindly placed at our disposal by the manufacturers, we find ourselves in a position of some difficulty. It is a pleasant kind of difficulty —that known as *l'embarras de richesse*—and the only way to meet it is (if we may be allowed to use another quotation) to "put down one's foot with a firm hand," and keep rigorously within the lines of our title, The *Portable* Steam Engine.

Of the allied types, semi-fixed, semi-portable, under-type, and the rest, there is more than sufficient matter before us to fill another volume. It is an obvious fact that every single one of the portable engines we have been examining, can, by the simple process of exchanging its carrying wheels for fixed supports, be converted into a *semi*-portable engine; but the converse does not hold good. Not every semi-portable can be converted into a portable engine, for the reason that these have grown in size and complexity to an extent far beyond the possibilities of road transport.

They have taken to themselves condensers (both of the "jet" and "surface" variety), necessarily accompanied by chimneys of 60 ft. or more in height, and sometimes by cooling towers—huge structures through which the heated condensing water is re-cooled for use over and over again—superheaters, oil-separators, and other appurtenances of the first-class modern stationary engine, whose place and functions they have to some extent usurped. With regret, we lay these aside, and proceed with the humbler task of considering the few remaining developments of the (strictly) portable steam engine.

SPECIAL TYPES

Fig. 74.—Messrs Robey's Compound Portable Engine, with Jet Condenser.

Fig. 75.—Messrs Ruston's Compound Portable Engine, with Jet Condensers.

SPECIAL TYPES

CONDENSING PORTABLE ENGINES.

When a condenser is used upon a portable engine, the exhaust steam, by which the necessary chimney draught is created, is diverted to the condenser through a three-way valve in the exhaust pipe, and the height of the chimney has to be increased to obtain the equivalent effect by natural draught. Usually 40 to 60 ft. is required, and this extra length is supplied by the makers, usually in 10-ft. sections.

Fig. 76.—Messrs Garrett's Jet Condenser for Portable and Semi-Portable Engines.

Messrs Robey & Co.'s compound condensing portable engine is shown in Fig. 74, and Messrs Ruston, Proctor & Co.'s in Fig. 75. A sectional view of Messrs Garrett's jet condenser is given in Fig. 76.

By the use of the jet condenser, the power of an engine may be increased by approximately 10 per cent., and the fuel consumption decreased by 15 per cent. (Engines should not be worked beyond the maximum powers stated by the makers.) The

Fig. 77.—Messrs Garrett's Superheated Steam Single-Cylinder Portable Engine.

SPECIAL TYPES

condenser can, of course, only be usefully applied where there is an abundant supply of water; the quantity required being 500 to 700 lbs. per brake horse-power per hour. The level of the water should not be more than about 15 ft. below the condenser.

SUPERHEATERS.

The only example before us of a superheater applied to a portable engine is that of Messrs Garrett, Fig. 77 (which shows, also, their improved feed-water heater attached). From an official trial carried out on 29th November 1907, by the editor of *The Engineer*, with one of Messrs Garrett's ordinary standard pattern engines taken from stock, as illustrated above, the following results were obtained (with steam at 170 lbs., 32 per cent. cut-off, at 200 revolutions per minute, a feed-water temperature of 146° Fahr., and a brake load of 26·3 H.P.), viz., steam per brake horse-power, 22·62 lbs.; coal (of 13,247 B.T.U.) per brake horse-power, 2·23 lbs.; water per lb. of coal, 10·1 lbs.; mean temperature of superheated steam, 510° Fahr.; mechanical efficiency, 89·7 per cent. Messrs Garrett remark that this result has, they believe, "never before been equalled by a single-cylinder non-condensing steam engine of this power."

While we are only able to give this one example of a portable engine fitted with superheater, we have reason to believe that several other leading manufacturers, making semi-portables for superheated steam, would place the smaller sizes upon wheels, if required, and thus convert them into really portable engines.

HIGH-PRESSURE SINGLE-CYLINDER PORTABLE ENGINES.

Of late years there has sprung up a class of single-cylinder engines constructed for pressures of steam hitherto associated only with compound engines. The engine illustrated above is Messrs Garrett's special "W.S." portable saturated steam engine, for working at a constant pressure of 140 lbs. The 20/24 B.H.P., and 25/30 B.H.P. sizes run at 200 revolutions per minute, and the 30/36 B.H.P. and 36/42 B.H.P. sizes at 180 revolutions. They are fitted with piston-valves, automatic crankshaft governor, chain lubrication crankshaft bearings, tubular feed-water heater, positive feed oil pump to cylinder, and forged steel balanced crankshaft. An official test with one of these engines developing 21 B.H.P. gave: steam per brake horse-power, 28·23 lbs.; coal (about 12,800 B.T.U.), 3·31 lbs.; mechanical efficiency, 91 per cent.; feed-water temperature, 165° Fahr.

Messrs Marshall, Sons & Co., Ltd., make their corresponding engine, "Class P," for a working pressure of 180 lbs., to run at speeds varying from 165 to 132 revolutions per minute, and in six sizes, from 18/24 B.H.P. to 45/55 B.H.P., with a "temporary maximum overload" of about 46 per cent. in excess of the ordinary maximum load. These engines, as shown in Fig. 79, are fitted with the Rider automatic expansion gear, already described, self-oiling dust-proof bearings, tubular feed-water heater, two steel tie-rods between cylinder and sliding crankshaft bearings; and forced lubrication to cylinder. Messrs Marshall state that the consumption of fuel in these engines is "reduced by about one-half to one-third the consumption of engines made a few years ago"—a statement which there is no reason to doubt, though it would seem to lack precision.

Messrs Ruston, Proctor & Co., Ltd., make their "Class 320" engines for a working pressure of 145 lbs. in seven sizes, from 18 to 53 B.H.P., running at speeds varying from 200 to 150 revolutions per minute. The Pickering governor is used, automatic oil-ring lubrication to the crank bearings, and forced lubrication to cylinder. Instead of a flat slide-valve a piston-valve is fitted, and, by a recent patent, the top of the firebox, at the tube-plate end, is hung up by stays to the steel flanged seating on which the cylinder is placed. The normal working load is about 80 per cent. of the powers given above, and can be obtained at a cut-off of about ·3 or a little more when working at 145 lbs. steam. We observe from the illustration that Messrs Ruston do not employ their steam-heated expanding stay between cylinder and crank bearings in these engines.

THE CIRCULAR FIREBOX BOILER.

We have before alluded to the difficulties attendant upon the transport of portable engines over rough roads—and, indeed, in the total absence of roads—when engaged upon their special duties upon the frontiers of civilisation.

Fig. 79.—Messrs Marshall's High-Pressure Single-Cylinder Portable Engine ("Class P").

Fig. 80.—Messrs Ruston's High-Pressure Single-Cylinder Portable Engine ("Class 320").

Fig. 81.—Messrs Brown & May's Portable Engine, with Circular Firebox.

Fig. 82.—Messrs Marshall's Portable Engine, with "Britannia" Boiler.

SPECIAL TYPES 103

The locomotive type boiler, though admirably adapted in every way as a steam generator for portable engines, has yet certain disadvantages for pioneer work, mainly proceeding from its low firebox. Even when the ashpan is removed from under it, the bottom of the firebox is of necessity very near the ground, as a glance at any of our illustrations will show.

Again, when the only available fuel—and in other respects a very suitable one—is composed of logs and branches of trees, there is manifestly waste of time in having to cut them up into short billets to suit the ordinary firebox. These are the considerations which have led most of the leading manufacturers to place upon the market a special type of boiler for rough work abroad. Their standard single-cylinder engines mounted upon boilers with cylindrical furnaces, and the usual tubes therefrom to the smoke-box, are illustrated in Figs 81 to 84 (it being understood that double-cylinder or compound engines can be supplied with these boilers if preferred).

Another adaptation of this type of boiler is illustrated in Fig. 85, which represents one of Messrs Robey's (compound) engines mounted on a circular firebox boiler with removable firebox and tubes.

As will be seen, the front plate of this boiler is attached by bolts and nuts instead of rivets. So also is the smoke-box tube plate. This modification (which can, we believe, be applied to their boilers by any of the firms above mentioned) enables the whole heating surface of the boiler—consisting of firebox, tubes, and smoke-box tube plate—to be drawn out in one piece for cleaning or repair. By the use of asbestos packing rings and closely pitched bolts all trouble from these huge steam joints is avoided, and it is possible for a gang of labourers to lay bare the whole interior of the boiler in two or three hours. The firebox, tubes, and boiler shell can thus quickly be chipped free from scale and replaced—an inestimable advantage in regions where skilled workmen are not easily obtainable.

This removable circular firebox and tubes is usually credited to German ingenuity; and most Continental portable and semi-portable engines up to very large sizes are so fitted. As a matter of fact, however, this boiler, removable firebox and all, is of British origin. Years before the prominent and respected German manu-

Fig. 83.—Messrs Robey's Portable Engine, with Circular Firebox.

SPECIAL TYPES

Fig. 84.—Messrs Ruston's Portable Engine, with Circular Firebox.

Fig. 85.—Messrs Robey's Circular Boiler, with Removable Firebox and Tubes.

SPECIAL TYPES

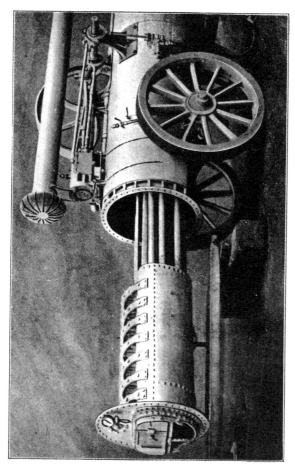

Fig. 86.—Biddell & Balk's Patent Removable-Tube Boiler, 1858.
(Messrs Ransomes & Sims, Ipswich.)

facturer, who claims to have introduced this type of boiler, went into business at all, a British patent (No. 620 of 1858) was granted to George Arthur Biddell and William Balk—two members of the staff of the then firm of Ransomes & Sims, of Ipswich—for the boiler illustrated in our Fig. 86, which boiler, without sensible alteration, is in widespread use at the present time: though not, curiously enough, by the firm (now Ransomes, Sims, & Jefferies, Ltd.) to whose enterprise it was originally due.

CHAPTER VI

PRACTICAL HINTS ON THE USE AND MANAGEMENT OF THE PORTABLE ENGINE

By describing what may be called the education of a portable engine—the various adjustments and attentions which are needed before a new engine is fitted to take its place as a useful and economical source of motive power—we hope to put into the hands of our readers the means of correcting incidental troubles which may arise in the course of daily working.

There are some items of knowledge which are the common property of the workmen actually engaged in the construction of, or in attendance upon, certain engines or machinery, which are not to be found in printed books, and seldom form the subject of papers addressed to technical audiences. But, in discussing the processes of the workshop, and the methods adopted by the man who actually carries out the ideas of others, we sometimes tap a vein of human insight and practical wisdom which is not beneath the attention of the most advanced student of mechanical science.

Now there is probably no greater nuisance known to users of steam power than the tendency manifested by some one or other of the bearings to "run hot," as it is termed, in spite of the possible fact that the bearing is amply large enough for its work, is not screwed up too tightly, that the lubricant is of excellent quality, and is supplied in quantity obviously more than sufficient.

Of course, if the bearing is defective in any of these four points, the remedy is indicated without going further; but sometimes, for no assignable cause within the knowledge of the steam user, the bearing cannot be kept cool, and much inconvenience is caused thereby.

The remedy is to ensure that a continuous and unbroken film

of oil is maintained between the working surfaces of the shaft or "journal," and its encircling bearing or "brass." If this film be preserved, obviously neither heating nor destructive wear can take place. The only problem is, how to preserve it?

Imagine for a moment, instead of the film of oil, a series of small cylindrical rollers interposed between the journal and the brass. If the shaft be now revolved the ring of rollers will rotate, as a whole, round the shaft at the speed of their own centres, which is, of course, midway between the peripheral speed of the shaft and the stationary brass, with both of which the rollers are in surface contact. The film of oil does exactly the same thing —it strikes the mean between the fixed and the revolving surfaces, and it will continue to do this unless the pressure is sufficient to squeeze it out from between them, or unless it be scraped or peeled off the shaft by the sharp edge of the brass, when, of course, contact ensues between the two metallic surfaces, and heating begins.

Again, a bearing in halves, when heated, does not expand away from the revolving journal and so tend to free itself. On the contrary, it closes upon the shaft, and so the heating proceeds in a compound ratio. A half brass which has been heated, moreover, does not regain its original diameter, but remains, even after being cooled, perceptibly tighter upon the shaft, and consequently a looser fit between the jaws of the plummer-block than it was originally.

The remedy for this is to scrape away the metal of the half brass from the edges downwards, so that only about one-fourth of the circumference, or $\cdot 7854 \times$ diameter, is in contact with the shaft. By this means the oil is led in instead of being peeled off, and the first step towards the maintenance of the continuous film is taken.

Some very interesting experiments have been made by Mr Dewrance bearing upon this subject, the results, shortly stated, being that for effective lubrication the oil must be fed in at the point of least pressure, and that a gradually decreasing space must exist between the surface of the brass and its shaft or journal, by which the oil may be, literally, induced to enter the actual bearing surface.

Herein he only demonstrates the theory of the well-known

practice of "easing away" the brasses at the sides. From this we deduce the fact that in bearings such as those of the connecting rods of horizontal engines, when the brasses are divided vertically, and the lubrication introduced at the top, no oil-channels are needed; while in the main bearings of the same engine, the oil should be introduced at the sides of the bearing, being brought thereto either, as he suggests, by external oil-pipes, or by grooves or channels cut into the internal surface of the upper half-bearing.

Heavy shafts, such as the main bearings of large horizontal or vertical engines, should be fitted with oil pumps for continuous feeding, while for quick revolution engines, forced lubrication under pressure possesses such important advantages that its use is becoming almost exclusively adopted for engines of this class. But, whatever system be adopted, the entrance of the oil at the point of least pressure is desirable.

There is one more important condition to be observed. In bearings which are fitted between fixed collars it must not be forgotten that a little end-play is necessary. The slightest pressure endwise upon such a bearing is sufficient to set it heating, and the expansion thus produced aggravates the evil, and would soon bring the engine to a standstill if not observed in time. These are the points requiring observation and correction in the case of new engines, so far as the bearings are concerned, and it is hoped that these hints, which experience in the trial running of new engines has suggested to the author, may be of service to users of engines whose bearings are inclined to give trouble.

A large and increasing proportion of the engines now built are fitted with cylindrical or bored-out guides; but in cases where two or four flat guide-bars are used, it becomes necessary to examine carefully the guiding flanges of the crosshead blocks, to make sure they are not too tightly fitted sideways, in which case, owing to the rapid motion of the crosshead, heating and cutting, or abrasion, would immediately set in. It is better to err upon the side of safety, and allow a little side-play.

Particular attention should be paid to seeing that the lubrication is directed to that surface of the guide-bar or bars upon which the pressure of the crosshead blocks, due to the angularity of the connecting rod, is exerted. If the engine be running "inwards"—that is, when the crank in the upper half of its revolution is running towards the cylinder—the top guide-bars are taking the pressure both on the inward and outward strokes of the piston; and the bottom bars if the engine be running "outwards."

A moment's consideration of the conditions will show that this must be the case; but, before leaving the subject of bearing pressures, we may put forward another proposition, perhaps more curious than useful,* which is far less generally recognised, and it is this: there is a portion of the peripheral surface of the crank journal which is never in the direct line of thrust of the connecting rod (even in a reversible engine), and where, theoretically, at any rate, a break in the continuity of the cylindrical surface would be productive of no harm whatever.

Consider a horizontal engine, working without compression. A little reflection will show that only one-half of the crankpin's surface is swept over by successive positions of the axis of the connecting rod. Imagine the crank to be on the dead centre nearest the cylinder, and the engine about to be turned slowly, say in the direction of the hands of a clock; then only the surface of the lower half of the pin will be utilised during the half revolution. Upon arrival at the other dead centre the thrust of the connecting rod is changed into a pull, and the same half of the pin comes into use again during the second half revolution.

Were the engine turning the other way round, the other half of the pin's surface would alone be in use, and of course in the case of a reversible engine the whole surface of the pin would, at one time or another, be in contact with the axis of the connecting rod. This is what would occur in an engine working without compression, or, in other words, when the reversal of the connecting rod stresses takes place exactly at the ends of the stroke. The effect of compression is, of course, to lead up gradually to pre-admission, or counter-pressure upon the piston, and so effect the reversal of stress without shock.

In every stroke there is a point at which the forces, whatever they may be, acting upon each side of the piston, are in equilibrium, as at A in the indicator diagram, Fig. 87, where the exhaust line is intersected by the compression curve. It is immaterial for our

* Communicated by the author to *Page's Magazine*, November 1904.

purpose at what part of the stroke this occurs, but assuming that equilibrium takes place at seven-eighths of the stroke, then, with a connecting rod four cranks in length, the effect upon the crank pin is as shown in Fig. 88.

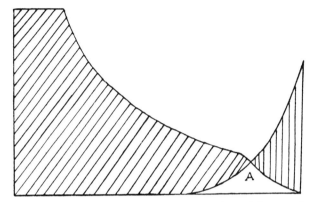

Fig. 87.—Indicator Diagram, illustrating the Effect of Compression.

It will be seen that with the proportions mentioned, there is a considerable part, about 80° of arc, of the crankpin's surface which is never used, in the sense of being in the direct line of connecting rod pressure. The crank, as illustrated, is supposed to be lying on the dead centre nearest to the cylinder.

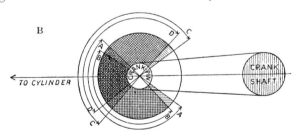

Fig. 88.—Diagram of Crank-Pin Pressures.

For a clockwise rotation, the arc A A contains all the points of connecting rod thrust, arc B B all the points of pulling effort. In the opposite, or anti-clockwise, direction, all the thrust of the connecting rod takes place within the arc C C, and all the pull within the arc D D. The part shown darkly shaded, from B to D, is covered twice in each revolution, whether the rotation be clockwise or the reverse.

To resume. It goes without saying that the guide-bars must be accurately in line with the axis of the cylinder, and that both must point exactly to the centre of the crankpin when on the dead centres. The axis of the crankshaft also must be truly at right angles with that of the cylinder, produced.

Undue wear of the brasses, combined with an obstinate "knock" in the bearings, may frequently be traced to a defect, original or acquired, in these particulars. An engine which is not "square" will never run satisfactorily; and the sooner it is put through its facings the better for all concerned.

This may be done in the following way: the back and front cylinder covers are removed, and a thin flat steel bar is fitted diametrically across the bore at each end. One of these bars has a V notch cut in it at the exact centre, the other being fitted with a small pointer of similar shape, exactly like the sights of a gun. The line of sight, if the cylinder be accurately placed on the boiler, will, of course, cut the exact centre of the crankpin. The centrality of the brass bush in the front cover of the bore of the stuffing box, and of the eye of the crosshead may be demonstrated in like manner by fitting a little sight-bar into each successively. If the crosshead be now moved back and forth in its guide-bars without apparent deviation of the notch, the engine may be considered true, so far.

The crank, which has been all this time set with its throw-journal, or crankpin, on the dead centre furthest from the cylinder, is now turned through half a revolution. If the centre of the crankpin still remains in the line of sight, it may be taken that the "squareness" of the main bearings, with respect to the axis of the engine, is established. But the truth of the crankpin (or throw-journal) itself remains to be proved. For this we shall require the connecting rod. This being put in place, with its large end cottered on the crankpin, but with its small end free, should be tried at both ends of the stroke by the crosshead (which we have already ascertained to travel accurately along the line of sight). If now the crank be placed in four successive positions, top, bottom, and both ends of the stroke, and the small end of the connecting rod drops fairly into its place between the jaws of the crosshead, we may conclude that the axis of the crankpin is parallel with that of the crankshaft.

All this being demonstrated to be in order, there is one more point to be established, and that is the circularity of the crankpin, which may be tested either by callipers in the ordinary way, or better, by placing the brasses upon it, enclosed in their strap, and cottered up just tightly enough to turn round without difficulty. If resistance to turning is uniform all round, the crankpin is all right, but if tight in one place and slack in another the source of the "knocking" is evident, and the pin requires to be re-turned in a lathe; or, if that be impracticable, to be very carefully "eased" on the tight places with a fine file, until the brasses can be turned round upon it without perceptible stiffness or slackness in places.

As may be judged from the space we have devoted to it, the crankpin is a very important detail, and if not absolutely correct in shape and position is the source of endless trouble to the unfortunate engine-driver.

Again, a "knock" in the engine may be caused through the tightness of the piston rings in the cylinder. The discussion of the various forms of piston ring in use is too large a question to enter upon here, but the ill effects of a tightly-fitting spring plug which has to be driven by the sorely-tried engine through four or five hundred feet of cylinder per minute may be easily imagined.

If there be reason to suspect a piston of being too tight, it may be put to the proof by laying the hand upon the connecting rod after the steam has been shut off, and while the engine is still running under the influence of the flywheel. The lag, or drag, of the piston will then be distinctly felt at each end of the stroke as a knock due to the reversal of the piston's motion. This is also the best time for detecting slack connecting rod brasses, as, owing to there being no compression or cushioning action of the steam, the change of motion in the reciprocating parts gives both audible and tangible evidence of any lost motion in their joints. Conversely, the instant of starting is the psychological moment for testing by touch the adjustment of the main, or crankshaft, bearings.

There are two meanings of the word "tight," as applied to a steam piston, one of which, as we have seen, is a vice, the other a virtue—in the sense of not being leaky. To test the "steam tightness" of the slide-valves and pistons, we may proceed as follows: Set the crank on one of the dead centres (in the case of a single-cylinder engine) and turn the steam on suddenly. If the exhaust nozzle be accessible the hand should be placed just over the orifice. No steam will issue if slide-valve and piston are in order. If, on the contrary, one or both are leaky, the fact will be unmistakably impressed upon the observer. (It may be remarked, in passing, that it is preferable to use someone else's hand in this experiment.) The engine should be tested upon both dead centres in this way.

If there be a "blow-through" at the exhaust nozzle, it may proceed either from the piston or from the slide-valve. To distinguish between these two possible sources of leakage, the crank is placed at about 30° from the dead centre (in which position the steam ports are all covered), and steam is turned on. If no steam passes now, the leakage may be confidently ascribed to the piston.

Piston rings which have worn "slack" in the cylinder may, as a temporary measure, be sprung out again by gently tapping the inner surface with a light hand hammer upon an anvil or some hard and heavy flat surface. This has the effect of expanding or "stretching" the inner circumferential surface and so opening the ring out. Care should be taken to remove any burrs upon the edges of the rings by the use of a fine file. It may not be superfluous to remind the operator that cast-iron piston rings are easily broken, without much effect upon the surface skin, by the use of too heavy a hammer.

Many slide-valves, from their form, have a tendency to arch, or "buckle," themselves when hot, so that a valve which is a perfect fit all over its surface when cold will bear only upon its two ends when heated up to its working temperature. So well is this understood, that it is a common practice to allow for it, in surfacing the valve, by scraping the ends so that they do not touch the surface plate when cold.

In a double-cylinder engine, if fitted with reversible eccentric sheaves, the simplest way of testing the valves and pistons is to set one of the valve eccentrics to "run backwards." The engine is then "hove-to," as a sailor would say, and can neither turn one way nor the other, in which condition it is, of course, perfectly tractable, and may be placed in almost any required position, and the steam turned on, for valve testing and observation.

In a compound engine the high-pressure piston and valves may be tested by the pressure gauge (if there be one fitted, as there should be) on the receiver, *i.e.*, the low-pressure steam chest. This will show at once if pressure is accumulating there through leakage in the high-pressure valves or piston. The low-pressure side, in its turn, may be set on its dead centres and steam admitted direct by the bye-pass valve to the low-pressure steam chest, from which it will escape through the exhaust pipe to the atmosphere in case of leakage through valve or piston.

We now come to the "setting," or adjustment, of the slide-valve. The four "beats" or puffs of steam for each revolution of the wheels of a locomotive engine are pleasing to the accustomed ear; but, if "out of beat," produce a sensation of lameness—a horrid discord to the engineer. Similarly, with a portable engine. It is such an easy matter to set the valves that there is no excuse for the driver who is content to let his engine beat "in dots and dashes," without an effort to put it right.

In engines with simple slide-valves, whether with one or two cylinders, the valves can be set by observing the beats of the exhaust steam with a sufficient approximation to accuracy. For instance, if the heaviest beat occurs when the crank is furthest from the cylinder, there is too much steam admitted at the back end of the cylinder, and the valve spindle must be adjusted by lengthening it slightly, there being usually a screw adjustment for this purpose. If too much steam is being admitted at the front end, and the heavy beat is heard when the crank is nearest the cylinder, then the valve spindle needs shortening. A driver, with a well-attuned ear, can set a valve in a minute or two in this way, so that an indicator diagram taken afterwards will show an almost exact balancing of the two sides of the piston.

If there be an expansion valve working on the back of the main slide-valve, and no indicator be at hand, the valves may be adjusted—provided the eccentrics are properly placed—by temporarily removing the expansion valve, and setting the main valve "by the lead," that is, adjusting the main valve spindle for length by observing the positions of the valve when the crank is on the dead centres, and equalising the lead opening for each end of the cylinder. The expansion valve being then put in place, and the cover replaced on the steam chest, the engine may be started, and the expansion valve set "by ear" as above described. Of course, in the case of a new engine the indicator

Fig. 89.—Automatic Expansion Gear (Messrs Clayton & Shuttleworth, Ltd.).

is always used for valve setting, if there be anything beyond a plain slide-valve to be dealt with.

Where the engine (being provided with an expansion valve) is also supplied with automatic expansion gear, the latter commonly consists of a swinging link worked by the expansion eccentric. The radius rod from the expansion valve spindle is pivoted to a die capable of sliding in the curved slot of the link from top to bottom, and the point of cut-off is determined by the height of the die in the link, as fixed by the position, at the moment, of the governor sleeve, to which it is connected by a suspension rod and lever, Fig. 89.

There is an apparent difficulty in connection with this system, owing to the error in the piston's movement caused by the angularity of the connecting rod, which may be worth clearing up. It is the practice in designing valve gears, as a rule, to assume the connecting rod of infinite length; in other words, to consider the piston's movement, in its relation to the crank, as alike at the two ends of the stroke.

In Fig. 90 the two halves of the crank's revolution (supposed to be divided by a vertical diameter, C O D, of the crankpin circle)

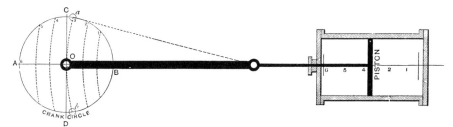

Fig. 90.—Effect of Connecting Rod upon the Position of the Piston.

and the positions a and b of the crankpin, when the piston is in the middle of the stroke, are shown.

It is evident that when the crankpin occupies the positions C or D, the piston will be drawn forward. Hence assuming uniform rotation of the flywheel, the mean velocity of the piston will be greater while the crank is performing the half revolution next the cylinder, than during the corresponding half furthest away from it. If the connecting rod is, say, five cranks long, the piston's velocity just before and after the beginning of the stroke at the end furthest from the crankshaft is 50 per cent. greater than in the corresponding positions at the other end, these differences rapidly diminishing as the stroke proceeds, until, at near midstroke, they, of course, disappear.

The problem is to get equal and symmetrical indicator

diagrams from both ends of the cylinder, at all points of cut-off, under these dissimilar conditions.

Fig. 91 (taken from an engine fitted with Robey automatic expansion gear) shows that this can be done, and we have no doubt that the same effect can be produced by any of the auto-expansion gears already illustrated. Let us see if we can unravel this rather knotty problem.

Now, putting aside the complexities of the automatic gear for a moment, suppose a simple slide-valve with an eccentric adjustable for different points of cut-off, but giving a constant lead. We lay an ordinary spirit-level (such as every erector carries in his pocket) upon one of the guide bars of the engine, and note the position

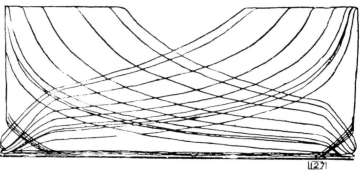

Fig. 91.—Indicator Diagrams, showing Equal Distribution at All Points of the Cut-off.

of the air bubble, then laying the spirit-level upon the large end strap of the connecting rod, we turn the crank until the bubble is exactly at the same spot again. The engine will then be upon one of the dead centres.

If, now, we turn the engine round to each of the dead centres successively, and, with the steam chest cover off, adjust the length of the valve spindle until the lead (or amount the steam port is open at the dead centre) is alike for each end of the stroke, we shall find that the point of cut-off is anything but equal for the two ends. If the engine were set to work with the valve in that position, the beats of the exhaust (which to the practised ear afford an excellent measure of the quantity of steam admitted respectively to the two ends of the cylinder) would be wildly irregular. We shall find that the valve spindle will have to be adjusted by unscrewing (which can generally be done while the engine is in motion) so as to move the slide-valve further away from the crank; a little only, if for an early point of cut-off, but a good deal for a late cut-off, and so, *pro rata*, for all points in the stroke. Now, the engine having been attuned to the sensitive ear of the operator, we shut off the steam, open up the steam chest again, and see where the slide-valve is. Putting the crank on the outer dead centre we are astonished to find the steam port nearest the crank perhaps one-third open, while its fellow at the other end, with the engine on the other dead centre, is hardly open at all. As already mentioned, the discrepancy between these two openings is greater

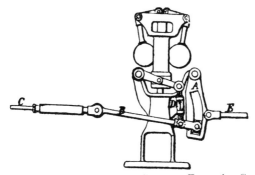

Fig. 92.—Sketch of Automatic Governor Expansion Gear
(Messrs Robey & Co., Ltd.).

or less, according to the fraction of the stroke at which steam is cut off, and, in the case of automatic expansion gear, where the cut-off point is continually varying, we are in effect called upon to be continually altering the length of the expansion valve spindle, if we wish to secure symmetrical indicator cards for all points of cut-off within range of the gear.

How is this rather difficult condition fulfilled in practice? Referring to the diagrammatic sketch, Fig. 92, the expansion link A is left by the erector set so as to swing through equal arcs on each side of a vertical line dropped through its centre of suspension; and if indicator diagrams be taken under these conditions, it will be found that though a pair of symmetrical diagrams can be got for any one cut-off point by adjusting the length of the expansion valve spindle C, or of the radius rod B, yet they become unequal the moment the governor changes the position of the die in the link A.

The remedy, and an effectual one, is found in a lengthening of the expansion eccentric rod E, to an amount found by trial when taking the indicator cards. (In Fig. 89 it will be seen that Messrs Clayton & Shuttleworth have provided a means of doing this by a right and left handed adjusting socket. As it is an adjustment made once for all, this would seem rather a doubtful benefit, as leading to possible misuse.)

The effect of this lengthening, by whatever means it is accomplished, is to cause the expansion link to swing or vibrate through a longer arc on the side nearest the cylinder (or in other words, to incline the link in that direction), with the result that, as the radius rod B is raised by the governor, not only is the travel of the valve reduced, giving an earlier cut-off, but the virtual length of B is reduced also, in corresponding degree; and *vice versâ*. We are thus able to obtain symmetrical diagrams from the two ends of the cylinder at any and every point of cut-off.

Before leaving the governor, we must note that it is a *sine qua non* that the expansion valve be set to cut off at zero when the governor is at its highest position, otherwise accurate governing cannot be secured. The governor itself will also need a little attention. Without going into theory, we will just put ourselves into the position of the man who has to adjust the governor so that it is stable without being sluggish, and sensitive without being unsteady.

In any spring-loaded centrifugal governor, whether its function be to control the movement of the expansion valve, or simply to regulate the amount the throttle-valve is open, the first essential is that when the balls are out to their fullest extent, no steam shall be admitted to the cylinder, or, at any rate, none sufficient to move the engine round. This being made sure of, either by adjusting the relative positions of governor and throttle-valve, or of governor and height of die in the expansion link, we next have to adjust the strength of the spring (usually in compression) by which the centrifugal force of the balls is more or less balanced.

Now, there is a particular degree of compression in a governor spring which may be termed the critical point of the governor under consideration; and it is, in practice, usually found by trial—by experimentally compressing the spring to the point at which its rate of increase will just keep pace with the increasing power of the balls to compress it; a suitable spring, adapted to the conditions, as determined by calculation and previous experience, being, of course, assumed.

At this "critical point" the governor will "hunt," as it is called—the centrifugal force being balanced by spring pressure at all points, a difference of one revolution more, or one less, in the engine will cause the balls to fly out or in to the limits of their range; a most unhappy state of things for the engine, which embarks upon a series of short runs, as if possessed by a fiend.

By reducing cautiously the pressure upon the spring, through the means provided for its adjustment, we gradually arrive at a point of extreme fineness of governing—unless this is prevented by friction, or "sticking" in the governor or its gear. Unless every joint in the details of the gear partaking of the movement of the governor sleeve be perfectly free, the governor will "hang" at some point in its movement until the engine's speed is increased or reduced sufficiently to overcome the resistance. This is easily distinguishable from true "hunting," though, at first sight, the effect upon the engine is very much the same.

Supposing all frictional resistance of this kind to be eliminated, it should be possible, upon throwing off the load, for the governor to settle down, after a few gentle oscillations, to a speed only 1 or 2 per cent. above the normal. This, however, is largely a matter of flywheel. If the energy in the flywheel be insufficient to absorb the fluctuation in the turning effort at the crankpin, within, say, 2 per cent., it is manifestly waste of time to try and control the rate of revolution within this amount unless we desire to see every stroke of the engine faithfully reproduced in movements of the governor sleeve. The contrivance known as the dashpot—consisting of a small cylinder filled with oil, and having a slightly leaky piston, past which the oil must squeeze its way—is sometimes used to correct, in a way, a deficiency in flywheel power. This has a damping effect upon a too lively governor, but it is in no sense a substitute for a flywheel of proper weight.

It will be remembered that we began our adjustment by making the governor "hunt" (which is a sort of datum-point from which our experiments radiate), and that we gradually reduced the pressure on the spring until we arrived at a degree of compression which gave, in the absence of detrimental friction, an almost exactly poised but steady governor. If, now, we continue to reduce the spring pressure, we lose in sensitiveness but gain stability, reducing it, so to speak, from the conditions of a laboratory balance to those of a grocer's scales. The governor will now not be affected by small changes in the speed, but, on the other hand, when it responds to a change of speed sufficient to influence it at all, it will move to the required point without those preliminary oscillations which before characterised its movement.

These relaxations of the spring pressure, besides reducing the sensitiveness of the governor, have the effect also of reducing the rate of revolution of the engine, and should this have fallen below the required speed of rotation, either a new and slightly stronger spring must be put in, or the ratio of the belt pulleys in the governor drive will have to be modified to suit. Should the speed of the engine's rotation be too high when the spring has been adjusted to give the best governing results, the latter may be reduced in strength as required by carefully grinding the coils externally so as to reduce their sectional area.

A few remarks upon the management of the fire may not be out of place, as, in order to work the engine with any degree of economy, it is necessary that the firing should be conducted on a proper system. A careless or ignorant fireman can easily neutralise the effect of all the improvements ever devised by the makers of portable engines.

Theory and successful practice unite in proclaiming that a high and continuously maintained furnace temperature is essential to proper combustion. No one who has enjoyed the opportunity of watching the firing upon an express locomotive engine during a long run can have failed to notice the extraordinary efficiency of pure radiant heat. For many miles before the conclusion of the run the stoker has ceased to ply the furnace with coals. The fire, urged by the tremendous power of the blast, becomes a glowing white-hot mass, and burns down thinner and thinner, the engine, meanwhile, steaming with the utmost freedom. Mile after mile this goes on, until the lights of the terminal station are in sight, and the engine brings up its heavy train with barely as much incandescent fuel in the firebox as suffices to cover the bars. Under these conditions, evidently, the furnace temperature is at its highest, and that of the chimney at its lowest, hence the ideal state of working through the maximum range of temperature is approached—for a time.

The total efficiency of the whole apparatus of the boiler (which necessarily includes the personal efficiency of the fireman) is a fraction formed by dividing the actual quantity of water evaporated by the evaporative value of the fuel. Thus, in a particularly fine result (not, it is true, in a locomotive type boiler) recently recorded we have an evaporation per lb. of fuel of 9·95 lbs. at 160 lbs. pressure. The ascertained value of the coal being 14,049 B.T.U. (British Thermal Units) per lb., and the total heat of the steam, calculated as from 66·5° Fahr., being 1160·5 B.T.U., we have

$$\frac{1160·5 \times 9·95}{14,049} = 82·2,$$

and this figure, 82·2, is the percentage of total efficiency of the boiler. With the superheater in use, 154° Fahr. above the steam heat, the total efficiency amounted to 87·46 per cent. This is what we have to aim at; but how many steam users are content, year in and year out, with a rate of evaporation of little more than half this? where the fireman, with an energy which would be praiseworthy if applied, for instance, to the loading up of a cart, shovels in the coal more or less at random. The fireman, whether he knows it or not, is a manufacturing chemist on rather a large scale, and it rests entirely with himself whether he manufactures the article his employer wants, or something very different, by a slight modification of his methods.

The locomotive firebox, with its roomy proportions as compared with the area of the grate, is an excellent furnace; and when the advantages of the firebrick arch and the ventilated firedoor with deflecting plate are combined with it, excellent results are possible with skilled firing. A consideration of the following facts may be of service to the ambitious fireman, if stated in a way which will be intelligible to him.

In an analysis of a sample of engine coal, the constituents were found to be approximately thus:—Carbon, 75 per cent.; hydrogen, 4·25 per cent.; nitrogen, 1·5 per cent.; oxygen, 10·75 per cent.; sulphur, 0·25 per cent.; incombustible matter, 8·25 per cent. The carbon, constituting three-fourths of the whole, is the principal agent in the production of heat, and it combines with the oxygen of the air entering the furnace in one of two ways: either it combines with twice its weight of oxygen, and forms carbonic acid (CO_2), or it unites with its own weight of oxygen, and forms carbonic oxide (CO).

Now we are coming to the point. For convenience, suppose that instead of dealing with $\frac{3}{4}$ lb. of carbon (*i.e.*, 1 lb. of coal), we take the carbon itself as 1 lb. 1 lb. of carbon, in burning to carbonic acid, liberates, in round numbers, 14,500 heat units (B.T.U.); the same lb. of carbon, burning to carbonic oxide, yields only 4,450 heat units. Hence coal, or carbon, may be burnt under two conditions, in one of which it is just three and a quarter times as valuable as in the other; or, to put it in a form readily grasped, 1 lb. of carbon (representing $1\frac{1}{3}$ lbs. of coal) may either be capable of evaporating, ideally, 15 lbs. of water, from and at 212° Fahr., or it may only be capable of evaporating 4·6 lbs. It is entirely a question of firebox temperature and the judicious admission of air above the fire through the ventilated firedoors.

The golden rule for firing successfully, and with economy, may be stated in two lines:—*The fire must be very thin and very bright, and the coal must be put on in small quantities and often.*

The thickness of the fire will vary so much in different engines and under different conditions that it is not possible to lay down a hard and fast rule for the thickness, which only experience can determine, but *the thinnest fire with which the engine can be made to keep steam is the right one.*

Again, if the firebox is not fitted with a firebrick arch, the "pull" of the blast will be greatest through the burning fuel nearest to the tube plate, and gradually lessen to its minimum at the back plate under the firedoor. The fire, therefore, should be thicker next the tube plate, sloping down to the back-plate, to maintain a uniform brightness. If the fire is collected in a heap in the middle of the grate, a stream of comparatively cold air is continually passing up all round the walls of the firebox, which is highly injurious to it, and the firebars also suffer from the additional weight imposed upon them at their weakest point, the centre.

When the fire is just burnt bright, the steam pressure will show signs of rising. A new thin layer of coals will check this for a few moments, and then the firebox will be filled with flame, and the pressure will look up again. By means of the ashpan damper, it ought to be possible to keep the pressure for hours at a time without its varying more than a couple of lbs. in either direction. The fire should never be allowed to burn into holes, nor should it be tortured about with a rake. These are the resources of a bad fireman who has allowed his air spaces to become clogged with clinker. The shovel and the damper are the only tools necessary in ordinary running.

Ashes should not be allowed to accumulate in the ashpan, and so choke up the engine's breathing space. It is very easy to get a second fire alight in the ashpan, burning up the air before it passes into the furnace, and causing the firebars to droop because they cannot be kept cool.

The firing system of a locomotive type boiler is totally different from that of a Lancashire boiler, with its long and narrow grates, and (by comparison) enormous reserve of heated water. The fire in the loco. type is soon up and soon down, and the effect of feeding it with large quantities of coal at a time, and allowing the fire to burn down low before replenishing it, is to lower the furnace temperature by the abstraction of the heat required to gasify the volatile constituents of the new supply. This lowering of the temperature occurs at the exact time when the highest temperature and the greatest admission of air are required to effect the combustion of the hydrogen, which as a consequence passes away unconsumed, having added nothing to the useful heat of the furnace.

A portion of the carbon, too, which at the moment of throwing on fresh coal was floating about in the furnace at a high temperature in search of oxygen to combine with, is cooled down below the temperature necessary to its combustion, and passes wasted away into the atmosphere, either in the form of smoke, or combined with the hydrogen as olefiant gas (a valuable compound of six parts by weight of carbon to one of hydrogen).

It rests largely with the fireman whether the furnace produce the beneficial CO_2 or the depraved CO, but in avoiding Scylla, he may fall into Charybdis. In his anxiety to produce the desired effect, he may admit far too much air, in which case he is lavishing his heat units in heating up large volumes of air, and thereby lowering the temperature to a point at which complete combustion cannot take place. Then his carbon goes up the chimney, heat units and all, and advertises the fact of his unskilfulness to the surrounding neighbourhood by a banner of dense black smoke.

All this is, of course, the merest commonplace—in a sense—but we have endeavoured to extricate the salient facts from the mathematical and chemical formulæ into which they are generally frozen, so that he who runs may read.

Now, a word as to the care of the boiler itself. It is an imperative necessity, if the engine is to be kept out of the repairing shop, that the boiler be kept perfectly clean, both inside and out. A boiler may be destroyed in two ways: it may be torn to pieces by an explosion, or it may be done to death just as effectually, but more quietly, by the slower operation of corrosion; the firebox may be ruined by overheating caused by dirt and scale filling up the water spaces around it, and thus keeping the water away from the plates—which for want of it become red-hot and unable to resist the pressure inside, and consequently bulge out between the stays like a cushion and crack.

Upon the other side of the same plates—the fire side—very probably the process of destruction is going on also, in this way. The driver, we will say, has discovered that by throwing water into the firebox, when the fire is all but out, the clinker (or incombustible matter which adheres to the firebars) by its sudden contraction, will of itself become detached without the trouble of breaking it away by the "slice," or chisel-bar. The wet ashes remain all night in contact with the firebox plates, and not only is corrosion set up, but the space left for the expansion of the firebars all round the box becomes filled up with a solidified mass, as hard as iron, which causes the daily expansion to become a pressure upon the plates. Thus tortured, is it to be wondered at that the unfortunate firebox, overheated, corroded, distorted, does not last out half its days?

No leakage from any mud-hole, especially from the one in the smoke-box, should be allowed to continue beyond the next blowing-off day. A continual oozing from any hole or joint about the boiler will, in a surprisingly short time, eat away the plate all round the leaky place. In engines fitted with the old-fashioned cast-iron saddle, an unsuspected saddle-bolt hole may continue to leak for months or years, underneath the lagging of the boiler barrel, with results disastrous to the boiler plate. The lagging should be removed about the saddle occasionally to make sure that all is right.

Occasionally a firebox stay-bolt may show signs of leakage, which should be stopped at once by caulking round the head. Once a week the steam should be blown off by the safety-valve, after the fire has been drawn (never by running the engine without a fire, or the tubes will begin to leak) and the water allowed to escape through one or more of the bottom mud-holes, which will have the effect of carrying off a good deal of loose scale and preventing its deposit on the plates. Every ten to twenty-one days, according to the class of water used and the number of hours worked, the interior surfaces of the firebox should be gone over with a scraper, and the loosened scale removed. A jet of water from a hand force-pump, or otherwise, is useful in washing off deposit from places inaccessible to the scraper. The interior surface of the tube plate between and among the tubes should receive particular attention in this way. Finally, if the engine is engaged in work of a regular character, where the power absorbed is pretty constant, a careful log or record of the coal consumed should be kept, and any recorded increase noted, and the cause searched for.

These observations, which have already far exceeded the limits originally intended, could easily be carried very much further, as many omissions occur to the writer in looking over these pages. But most of the important features in the proper maintenance of the portable engine have been touched upon, and we must pass on to our next subject, the slide-valve and its action.

CHAPTER VII

THE SLIDE-VALVE AND ITS ACTION

IN the earlier editions of this book a good deal of space was occupied in explaining the elementary details of the steam engine. It is hardly necessary to include this matter in the year 1911, and we may assume that the reader is conversant with the ordinary slide-valve, so far as a knowledge of how the steam is alternately admitted to and exhausted from the two ends of the cylinder.

Fig. 93 shows the cylinder in section with the slide-valve in

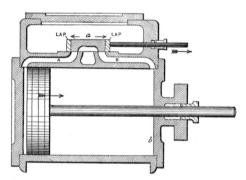

Fig. 93.—Section of Cylinder and Steam-Chest.

its central position. The eccentric sheave for actuating a valve like this (that is to say, whose length a equals exactly the distance between the outer edges of the steam ports) would be placed at right angles with the crank, and would have a radius or "throw" equal to the width of the port, Fig. 94.

In Fig. 93 both steam ports are exactly closed by the valve, but the slightest movement of the crank in either direction would uncover one port or the other. The port would be wide open when

THE SLIDE-VALVE AND ITS ACTION 131

the crank was at half stroke, and would close at the end of the stroke. The effect of this would be, of course, that a whole cylinderful of steam would be admitted twice per revolution. This is working without expansion, and early slide-valves were made in this way.

It was soon discovered, however, that this was an unnecessarily wasteful method, and that it was only required to increase the length of the slide-valve and advance the eccentric somewhat with

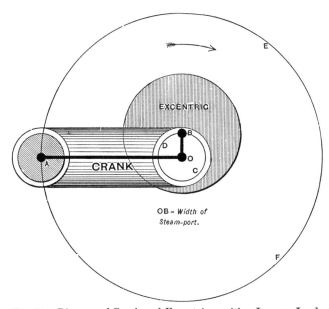

Fig. 94.—Diagram of Crank and Eccentric : neither Lap nor Lead.

regard to the crank, to take advantage of the expansive property of the steam.

Fig. 94 shows the crank arm and eccentric sheave for a lapless or non-expansive valve. The distance O A represents the crank radius, and O B the radius of the eccentric, equal, of course, to half the stroke, or travel, of the piston and slide-valve respectively. The eccentric, it must be remembered, is only a crank with an exaggerated diameter of crankpin, so large relatively to the radius or throw as to include both centres within its diameter. In succeeding diagrams like Fig. 94 only the black radius lines

O A, O B will be shown, and the actual representations of the crank and eccentric omitted. Now, before we proceed with the designing of a real slide-valve there is one more point to be considered.

If the reader will turn back to Fig. 93 he will see that whereas the black valve with no lap is just upon the point of opening the steam port A by the slightest movement in the direction of the arrow, the new valve, as we may call it—represented by the addition of the shaded *lap* portion at each end—would have to be

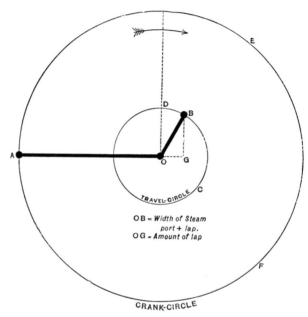

Fig. 95.—Diagram of Crank and Eccentric, with Lap.

moved forwards the amount of the lap before it would be ready to open the port. The piston is at that end of the cylinder, ready to begin its stroke, and the slide-valve must be in a position to admit steam to it on the instant. The eccentric sheave must, therefore, be so placed in relation to the crank that every time the piston arrives at either end of the cylinder (in reality a little sooner, for reasons to be explained presently) the edge of the slide-valve at that end shall be upon the point of uncovering the steam port to supply steam for the next stroke.

THE SLIDE-VALVE AND ITS ACTION

All that is required to fulfil this condition is a shifting of the eccentric further round upon the crankshaft to an extent sufficient to move the valve forward the width of the lap.

Fig. 95 shows the alteration in the eccentric. The radius O B has been increased by the amount of the lap (being now the width of the port plus the lap), and it has been moved round upon the shaft until a perpendicular, let fall from the centre of the eccentric B, falls at a distance O G, equal to the lap, from the centre of the

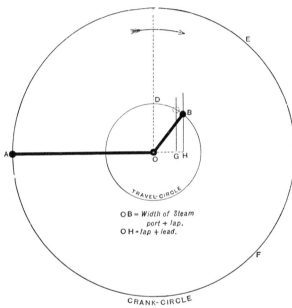

Fig. 96.—Diagram of Crank and Eccentric, with Lap and Lead.

crankshaft O. The angle D O B, by which the angle A O B is in excess of a right angle, is called the angular advance.

As now arranged, our engine is constructed to take steam at the moment the crank is passing the dead centre, or in other words, when the piston is at either end of its stroke. But the piston, crosshead, and connecting rod are heavy, and the sudden reversal of the motion of these parts, some two hundred and fifty times a minute, produces undesirable shocks and strains in the engine, so much so as to suggest the employment of some kind of spring or buffer at each end of the cylinder to take up the momentum of the moving parts. If we could so arrange that the slide-valve should begin to admit steam a little before the piston reached the end of its stroke, an elastic cushion of steam, the very thing we want, would be provided.

This can be done easily enough by giving the eccentric a little more angular advance, so that the port, when the crank is on the dead centre, shall be open, say, $\frac{1}{16}$ in., which will fulfil all requirements. This opening is called lead, or pre-admission, and the amount say $\frac{1}{16}$ in., is added to the lap in constructing a diagram as in Fig. 96.

Now, being conversant with the leading features of the slide-valve, we may proceed to construct, on paper, a valve such as would be used in a portable engine; and, in following it through one revolution of the engine, endeavour to make ourselves thoroughly familiar with the action of the steam in the cylinder.

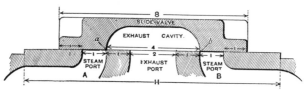

Fig. 97.—Section of Slide Valve and Cylinder Face.

The proportions we shall adopt are as follows, taking the width of the steam port as the unit:—

Width of steam port	1
,, exhaust port	2
,, bars, or spaces, between ports	1
Length of cylinder face	11
,, slide-valve	8
,, cavity in slide-valve	4
Travel of slide-valve	4
Lap of slide-valve	1
Lead of slide-valve	$\frac{1}{12}$
Exhaust lap	0

With these proportions the valve will cut off the steam at rather under three-fourths of the stroke. The cut-off will take place earlier if the travel be reduced, retaining the same lap, or if

THE SLIDE-VALVE AND ITS ACTION 135

the lap be increased, retaining the same travel; or it will take place later if these conditions be reversed. In each case the eccentric must be set so as to give the proper lead.

A useful formula for finding the point of cut-off, having the lap, lead, and travel given, is the following:—

Let L = the lap,
l = the lead,
S = stroke of piston,
T = travel of valve,
x = the point of cut-off.

Then, $$x = S\left[1 - \left(\frac{2L+l}{T}\right)^2\right].$$

Or, having the point of cut-off required and the lead and travel, to find the lap:—

$$L = \left(\tfrac{1}{2}T\sqrt{\frac{S-x}{S}}\right) - \tfrac{1}{2}l.$$

By way of example, we may apply this rule to the present case, taking the stroke of the piston as 12 in. and the lap as 1 in., with lead and travel in the proportions given above. The travel will then be 4 in., and the lead $\frac{1}{12}$ in., or ·0833 in.

x, or the point of cut-off in inches, will be:—

$$x = 12\left[1 - \left(\frac{2·0833}{4}\right)^2\right] = 8·75 \text{ in., or } 73 \text{ per cent.}$$

We now know, in the first place, that certain advantages result from an early cut-off of the steam in the cylinder; and secondly, that by giving lap to the slide-valve in a certain proportion to its travel, any desired cut-off point may be secured. (A simple slide-valve is not, however, adapted for a cut-off of less than about half-stroke, as the exhaust is seriously interfered with.) We have the means, by the formulæ just given, of determining what the lap should be for any desired cut-off, and, given the initial pressure on the piston, we can ascertain, by reference to a table of hyperbolic logarithms, the mean effective pressure on the piston for any desired point of cut-off thus: Let p = the mean pressure upon the piston in lbs. per square inch during the whole stroke; l = the fraction of the stroke during which steam is admitted; R the ratio of expansion (which may be expressed as $\frac{1}{l}$, or l inverted); H the hyp. log. of the fraction l in the table below; and P the pressure before steam is cut off.

Then,
$$p = P\frac{1+H}{R}.$$

We may give one example of the method of using this table. Let it be required to find the mean pressure in a cylinder where 60 lbs. steam is cut off at five-eighths of the stroke. Here the ratio R will be $\frac{8}{5}$, or 1·6, and the mean pressure

$$p = 60\frac{1\cdot 47}{1\cdot 6}, \text{ or } 55\cdot 1 \text{ lbs.}$$

TABLE OF HYPERBOLIC LOGARITHMS.

l, or the Fraction of the Stroke at which Steam is cut off.	R or $\frac{1}{l}$. The Ratio of Expansion.	H. The Hyperbolic Logarithm of R.
$\frac{1}{10}$	10	2·302
$\frac{1}{8}$	8	2·079
$\frac{2}{10}$	5	1·609
$\frac{1}{4}$	4	1·386
$\frac{3}{10}$	3·33	1·203
$\frac{3}{8}$	2·66	·978
$\frac{4}{10}$	2·5	·916
$\frac{1}{2}$	2	·693
$\frac{6}{10}$	1·66	·507
$\frac{5}{8}$	1·60	·470
$\frac{7}{10}$	1·43	·358
$\frac{3}{4}$	1·33	·285
$\frac{8}{10}$	1·25	·223
$\frac{7}{8}$	1·14	·131
$\frac{9}{10}$	1·11	·104

In any except the roughest calculations the pressure P should be expressed in lbs. above a vacuum (*i.e.*, 14·7 lbs. should be added to the gauge pressure), and the same amount should be subtracted from the result, which in the case given would make the result 53·96 lbs.

A German engineer, Professor Zeuner, has given us perhaps the most expressive graphic method of placing before the eye the movement of the slide-valve throughout the stroke as clearly (and much more usefully, because it can be measured) as if displayed upon the screen of a kinematograph. If the reader is not familiar with the Zeuner valve diagram we can promise that the few minutes' work involved in tracing out the valve movement by its means will be well rewarded. And it should be noted that not only can we construct the diagram from the conditions of a given valve, and find out exactly what it will do, but we can also lay down first the desired conditions of lead, cut-off, exhaust opening or closing, and compression, and, from the diagram, build up the proportions of a slide-valve which will fulfil them. We can see and criticise the whole movement from first to last with as much certainty as though we saw the valve at work in the completed engine.

The diagram is constructed as follows for an ordinary single slide-valve. A line A B, Fig. 98, representing to any convenient scale the stroke of the piston, is drawn as the diameter of a circle representing the path of the crankpin, called the crank circle. From the same centre is struck a smaller circle showing the path of the centre of the eccentric sheave. This is usually drawn full size, but for the sake of keeping this and succeeding diagrams within reasonable limits of size, we shall suppose our valve and cylinder face to be of the actual size of those in Fig. 97, and take $\frac{1}{4}$ in. as the unit. The travel of the valve being four times the unit, or 1 in., this will be the diameter of the travel circle. From the same centre o is struck, with a radius equal to the lap ($\frac{1}{4}$ in.), the lap circle. The crank is supposed to start from the point B, and to run in the direction of the arrows; A and B being thus the dead centres, or points in the revolution at which the piston is at the extremities of its stroke.

Lead, or pre-admission, begins, it will be remembered, before the crank arrives at the dead centre B, and the port is to be open one-twelfth of the unit, or $\frac{1}{48}$ in., when the crank gets to that point. To represent the lead, then, we draw the line o a, cutting the travel circle at b, $\frac{1}{48}$ in. below the centre line A B, and the point a upon the crank circle denotes the exact spot in the revolution of the crank at which steam admission begins.

Cutting this line oa at g upon the lap circle, and passing through the centre o, a fourth circle efg is described, whose radius equals one-half that of the travel circle. As explained upon p. 133, the radius of the eccentric equals the lap and the port opening together. The shaded portion of the circle efg, therefore, exterior to the lap circle, represents port opening. The two extremities of this crescent-shaped portion mark the points of opening and closure of the port respectively. As the

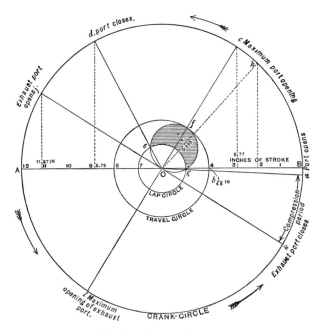

Fig. 98.—Zeuner Diagram.

radius oa gives the opening, so the radius od, drawn through the other intersection of the lap circle with the circle efg, gives at d the point where the valve closes the port and cuts off the steam. The radius oc, passing through the centre of the circle efg, points to the maximum port opening, $\frac{1}{4}$ in., and throughout the arc of steam admission, acd, the amount of port opening for any given position of the crank always equals the width of the shaded crescent at the radius corresponding. For example, the

THE SLIDE-VALVE AND ITS ACTION 139

port opening, when the crank is at the point h, is measured along the radius $o\,h$, and at that point equals ·235 in.

The line A B, representing the stroke of the piston, may be conveniently divided into equal parts representing inches of stroke. In this case it is divided into twelve parts, and a perpendicular let fall from the point d, where the cut-off takes place, cuts the line A B at a point 8·75 of these inches from the beginning of the stroke, thus confirming the correctness of the formula given on p. 135. A similar perpendicular from h shows that at 2 in. of the stroke the valve had uncovered the port to an extent of ·235 in. So the action of the valve, as it covers and uncloses the ports, can be closely followed throughout the stroke. As there is no exhaust lap (i.e., the inner edges of the exhaust cavity a and b, Fig. 97, coincide with those of the ports when the valve is at mid-travel) the point of exhaust opening j is distant just one-fourth of the circle, or 90° from the maximum steam port opening c, or at 11·07 in. of stroke. The exhaust remains open for half the revolution, attaining its maximum at l, a point diametrically opposite to c, and closes at k, from which point the steam yet remaining at atmospheric pressure in the cylinder is (as we shall see when considering the indicator diagram) gradually compressed by the advance of the piston to nearly admission pressure, and prepares the way for the admission at a of steam for a new stroke.

The portion of the revolution from k to a is called the compression period.

Thus the crank has performed, under our scrutiny, an entire revolution, but only so far as one end of the cylinder is concerned.* The points of cut-off and release differ slightly for the two ends of the cylinder on account of the angularity of the connecting rod, as will be explained presently. The diagram, however, indicates correctly the mean cut-off.

One more point may be noticed here. Temporarily assuming that the engine runs in the direction contrary to the arrows, and that the crank radius is lying at O A, then the angle A O c gives the correct position of the eccentric (whose centre would be at f) relatively with the crank.

* A complete diagram showing both ends of the cylinder, and corrected for the angularity of the connecting rod, is given at p. 143, Fig. 102.

In our consideration of the diagrams we have assumed certain proportions for the valve, and have constructed the diagram accordingly; but it is obvious that this procedure might be reversed, and the proportions of the valve, or of the travel, might be found by construction from a given cut-off and admission, laid down as points in the revolution.

Exhaust Lap.—To illustrate the nature of the influence exercised by exhaust lap upon the release and compression of the steam, we give another example of a slide-valve and its Zeuner diagram. The question of its desirability or otherwise will receive attention when we come to the consideration of the indicator diagram.

Fig. 99 shows a slide-valve and cylinder face exactly similar in all respects to that illustrated by Fig. 97, save that lap, to the extent of $\frac{1}{16}$ in., or one-fourth of the unit, has been added to the

Fig. 99.—Slide-Valve, with Exhaust Lap.

inner edges of the exhaust cavity. The exhaust lap appears on the diagram in the form of a small circle, struck from the general centre o, whose radius, $\frac{1}{16}$ in., equals the lap. From the point s is struck another circle pqr, equal and opposite to the circle efg;* and through the two points u and v, where the former cuts the little exhaust lap circle, are drawn the radii om, on. The point m in the crank-circle marks the opening of the exhaust, 11·425 in. from the beginning of the stroke, and n represents the closing point, distant $10\frac{5}{8}$ in. from the beginning of the return stroke. It will be observed that the effect of lap is to retard the opening, and accelerate the closing, of the exhaust port, and to prolong the period of compression, which has been increased from about $8\frac{1}{2}$ per cent. of the revolution to 14 per cent. The points j and k, which are the same as laid down upon the previous diagram, indicate the amount of alteration. Where so compara-

* This circle is the one which would be drawn to show the port opening for the return stroke in the same way as the circle efg was used for the out-stroke.

THE SLIDE-VALVE AND ITS ACTION

tively small a circle as that used here to denote the exhaust-lap is necessitated, it is difficult to show the exact point of intersection, but it must be borne in mind that these diagrams in actual practice are made at least full size, and may with advantage be drawn double full size, or even larger.

We have now seen how a Zeuner diagram is made useful in tracing out, step by step, the movements of the valve in relation to the crank, and how by its aid a slide-valve may be constructed

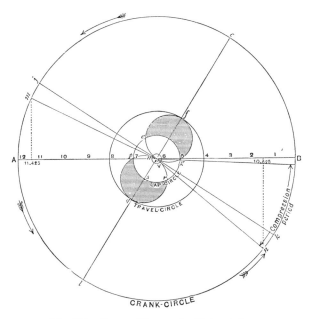

Fig. 100.—Zeuner Diagram for Exhaust Lap.

which shall perform certain operations at given periods. The diagram may be extended to show the action of the link motion reversing gear; of various forms of expansion gear, with or without separate cut-off valves; and of almost every form or modification of apparatus applied to steam engines for the inlet and release of steam. But it is not our purpose to follow it further here, or to enter into the geometrical reasoning which underlies the application of this form of valve diagram. The reader who desires to follow up this most interesting study will find the subject dealt with at length, and its various applications exhaustively discussed in Professor Zeuner's "Treatise,"* or in the author's work, "The Proportions and Movement of Slide-Valves."†

We have already drawn attention, to the irregularity in the piston's movement caused by the angularity of the connecting rod, and we reproduce Fig. 90 from that page for the purpose of showing how this may be accounted for in the Zeuner diagram.

Disturbing Influence of the Connecting Rod.—So far we have assumed, in diagrams Figs. 98 and 100, that the correct position of the piston for successive points in the revolution is always represented by an ordinate drawn from any point in the crank circle perpendicular to the line A B. Thus, in Fig. 98 we have taken it for granted that at h the piston has moved 2 in., 8¾ in. when

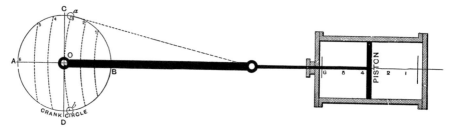

Fig. 101.—Diagram showing Irregularity induced by Angularity of Connecting Rod.

the crank reached d, and 11·07 in. at j. If the connecting rod were of infinite length this assumption would be correct. In practice connecting rods seldom reach more than six or seven times the length of the crank radius, and an irregularity is thus introduced into the motion of the piston to which we must devote a little space. In the skeleton diagram, Fig. 101, the connecting rod, represented by a thick line, is supposed to be uncoupled from the crank and lying in a straight line with the piston rod. The piston is in the middle of its stroke, and the connecting rod just reaches the centre of the crankshaft o. If now the rod be raised or lowered to the positions a or b upon the crank circle without moving the piston, it will be seen that the dotted arc $a\,o\,b$ described by the connecting rod cuts the crank circle into two

* Klein's translation (Spon).
† Technical Publishing Co., London.

THE SLIDE-VALVE AND ITS ACTION 143

unequal divisions, consequently that the centre of the piston's stroke does not coincide with the mid-positions of the crank, denoted by the perpendicular line C D passing through the centre O. Points in the crank circle corresponding to equal divisions in the stroke of the piston are, however, correctly transferred to the line A B, where, instead of straight ordinates, arcs with the connecting rod length as radius are employed.

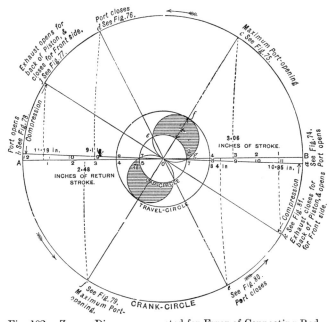

Fig. 102.—Zeuner Diagram, corrected for Error of Connecting Rod.

The Zeuner diagram, Fig. 102, is a reproduction of Fig. 98, save that the straight ordinates are replaced by arcs struck with a radius, representing the connecting rod, of seven times the length of the crank.* The diagram is also doubled to show the return stroke (in the lower semicircle) in order that the unequal points of cut-off, viz., 9·1 in. for the outward stroke, and 8·4 in. for the return, caused by the obliquity of the connecting rod may be clearly indicated.

* The length of the connecting rod in Fig. 101 is purposely reduced to four times the crank radius length in order somewhat to exaggerate the disturbing effect.

The previous diagram, Fig. 98, with straight ordinates, gives the mean of these two measurements, or 8·75 in., and similarly the piston, when the crank is at the two points j and k, is now shown to be distant from the corresponding ends of the cylinder

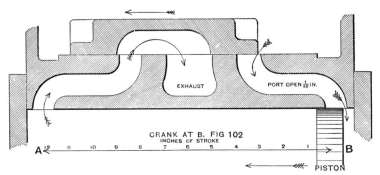

Fig. 103.—Slide-Valve Section.

11·19 in. and 10·95 in. respectively, the mean of these two measurements being that given upon the former diagram, 11·07 in.

As this diagram, Fig. 102, represents what is going on upon both sides of the piston simultaneously, it would perhaps be

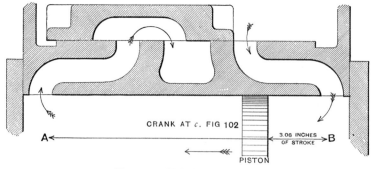

Fig. 104.—Slide-Valve Section.

advisable once more to follow the crank round the circle very briefly. The various positions of piston and slide-valve corresponding to the points in the crank circle B c d j, &c., are also shown in the series of sections given below, Figs. 103 to 110.

At B in the diagram (see Fig. 103) the crank is on the dead

THE SLIDE-VALVE AND ITS ACTION 145

centre, ready to begin its stroke. Steam is already being admitted behind the piston, the port being open $\frac{1}{48}$ in., and the front side is open to the atmosphere. With the crank at c in the diagram the

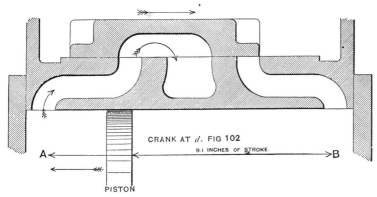

Fig. 105.—Slide-Valve Section.

maximum port opening, $\frac{1}{4}$ in., is reached (see Fig. 104), with the piston at 3·06 in. from its starting point.

Steam continues on until the point d is reached by the crank,

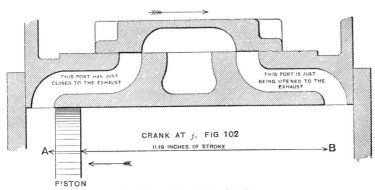

Fig. 106.—Slide-Valve Section.

when the valve closes, and the period of expansion begins with the piston at 9·1 in. of its stroke (Fig. 105).

All this time the front side of the piston has been in free communication with the atmosphere, but at the point j in the crank circle the slide-valve closes the exhaust port for the front

10

of the piston, and opens it to the back (Fig. 106), releasing the expanded steam and remaining open until the point k is reached half a revolution later.

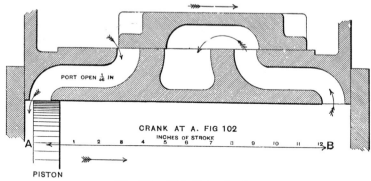

Fig. 107.—Slide-Valve Section.

This occurs at 11·19 in. of the piston's stroke. The period of compression now begins for the front end of the cylinder (supposed to have been filled with steam before our illustrative stroke commenced), and this continues until the crank reaches the point A in

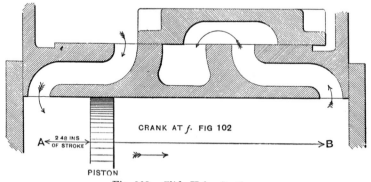

Fig. 108.—Slide-Valve Section.

the diagram (see Fig. 107). At A the port opens to the boiler, and admission begins. The maximum opening is reached at f in the crank circle (Fig. 108), with the piston 2·48 in. upon its return stroke.

At c in the diagram the port closes (Fig. 109) at 8·4 in. of piston, the steam expanding until the point k is reached, 10·95 in.

from the beginning of the stroke, where release begins (Fig. 110), and continues for half the revolution. At k also the valve regains its central position, the exhaust closes for the back end of the

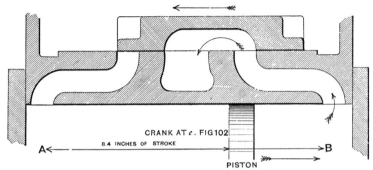

Fig. 109.—Slide-Valve Section.

cylinder, and compression of the small quantity of steam left in the cylinder takes place, until fresh steam from the boiler is admitted at a, and the cycle of operations goes on again, and continues as long as steam is supplied to the engine.

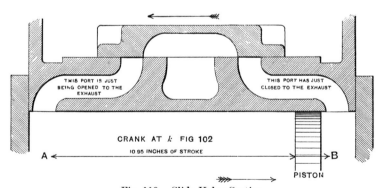

Fig. 110.—Slide-Valve Section.

Referring back to the diagram, Fig. 102, for a moment, it will be noticed that, as there is no exhaust lap, each end of the cylinder in turn remains open to the exhaust for half the revolution, and the valve closes for the one, and opens for the other, at the same instant.

THE PORTABLE STEAM ENGINE

Variable Expansion by Means of an Adjustable Eccentric.—
Hitherto we have considered only the action of an eccentric having
a fixed travel, and consequently an invariable cut-off point. The
variable expansion eccentric consists simply of a movable sheave
capable of being clamped in several different positions against a flat
disc, keyed to the crankshaft, and having a slot, or guide, formed

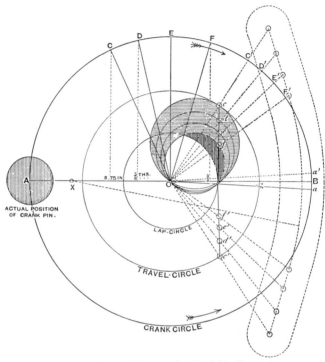

Fig. 111.—Zeuner Diagram for Variable Expansion.

in it, by which the position of the sheave for different cut-off
points is determined.

An inspection of the diagram of such an eccentric, Fig. 111,
will show, by the circles drawn cutting the lap circle at points
representing the different fractions of the stroke during which
steam is to be admitted, that the centre of the eccentric sheave
has only to be moved given distances in a straight line, from c,
which is the maximum cut-off, to b, when the admission is for

only three-eighths, to secure the closing of the port at those, or any intermediate, fractions of the stroke. The diagram, which is drawn double the size of the preceding ones for the sake of clearness, is thus constructed.

The fractions of the stroke measured along the line A B are transferred by ordinates to points in the crank circle C D E F, and radii are drawn from the centre O to these points. The amount of pre-admission, or lead, $\frac{1}{24}$ in., being measured upon the travel circle as before, the radius O a is drawn through it, and the angles C O a, D O a, E O a, and F O a are bisected. Circles drawn upon these bisecting lines c' d' e' f', cutting the lap circle at the radii C D, &c., and at the constant lead-point l, give by their diameters the four points representing c d e f, the positions of the centre of the eccentric sheave when set to cut off at the different fractions of the stroke, the crank being supposed at A.

The darkly-shaded crescent shows the port opening for the three-eighths cut-off, the lighter one that for the maximum, the intermediate circles the openings for the half and five-eighths respectively. The four points c d e f are situated in a straight line drawn at right angles with the lead radius o a. For reversing the direction of the engine's running, similar points, c' d' e' f', may be found upon the other side of the centre line A B; and the shape of the slot in the fixed disc, or expansion plate, by which the eccentric sheave is guided, is shown by the dotted lines outside the crank circle. As will be seen, the slot is formed by equal prolongations of the radii c' d' e' f' from the points c d e f, the two ends being connected by an arc, struck from the centre X.

The radii giving the points for opening and closing the exhaust are not shown in the diagrams, to prevent confusion, but they may be drawn in at right angles with the centre lines c' d' e' f' for each position.

It will be seen that the earlier the cut-off the earlier also is the closing of the exhaust, and at three-eighths of the stroke there is rather an excessive amount of compression. In engines constructed for working with higher degrees of expansion a separate cut-off valve is used which acts upon ports cut through the main valve, and so permits of alteration in the steam admission without affecting the exhaust closure.

CHAPTER VIII

THE INDICATOR DIAGRAM

HAVING constructed our slide-valve, now supposed to be running in the engine, we may proceed to ascertain its behaviour under steam. The indicator is the stethoscope of the engineer, and just as, by the aid of the Zeuner diagram, he is enabled to design a slide-valve calculated to produce in the cylinder of his engine certain effects of expansion and compression, so by the indicator diagram he sees, traced by the invisible steam itself, a record which to the initiated eye reveals in minutest detail the faults or excellencies of the apparatus he has made. Before the adaptation of this instrument to quick-running engines it is no figure of speech to say that we were completely in the dark as to the real behaviour of the steam in the cylinder. For instance, until the indicator enabled us to see that at high speeds pressure could not be maintained behind a piston unless the passages were large enough to allow the steam to get in and out quickly, we were without reliable information as to proper proportions of steam ports, and the importance of a free exhaust was more or less overlooked.

Consider the right-hand diagram in Fig. 112. It is constructed upon the assumption that the pressure varies inversely as the volume. For many reasons, which we cannot enter into here, this is only approximately true in the case of saturated steam, but at any rate we shall not err very much at present by acting as if the expression,
$$pv = \text{a constant},$$
were strictly correct in its application.

Premising that the length of the diagram represents the stroke of the piston, and that its height stands for pressure in lbs. per

THE INDICATOR DIAGRAM 151

square inch, we note the uniform height of the admission line
(right-hand diagram) at 55 lbs. pressure until the sixth inch of the

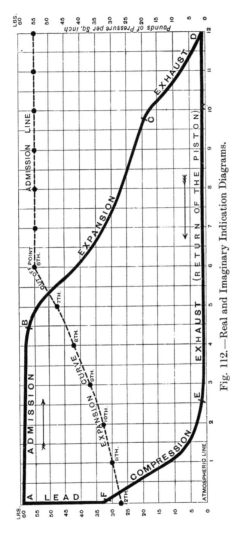

Fig. 112.—Real and Imaginary Indication Diagrams.

stroke is reached, from which point it descends in a curve terminat-
ing at $27\frac{1}{2}$ lbs. pressure at the twelfth inch. We may, if we
choose to take the trouble, find the points in this curve for each

successive inch of stroke by constructing a table of pressures thus:—

At the end of the first inch the pressure   -   -   -   = 55·00 lbs.
,, ,, second ,, ,,   -   -   -   = 55·00 ,,
,, ,, third ,, ,,   -   -   -   = 55·00 ,,
,, ,, fourth ,, ,,   -   -   -   = 55·00 ,,
,, ,, fifth ,, ,,   -   -   -   = 55·00 ,,
,, ,, sixth ,, ,,   -   -   -   = 55·00 ,,
,, ,, seventh inch, $\frac{6}{7}$ of 55, the pressure   = 47·14 ,,
,, ,, eighth ,, $\frac{6}{8}$,, ,,   = 41·25 ,,
,, ,, ninth ,, $\frac{6}{9}$,, ,,   = 36·66 ,,
,, ,, tenth ,, $\frac{6}{10}$,, ,,   = 33·00 ,,
,, ,, eleventh ,, $\frac{6}{11}$,, ,,   = 30·00 ,,
,, ,, twelfth ,, $\frac{6}{12}$,, ,,   = 27·50 ,,

Sum of pressures,   -   -   -   -   = 545·55 lbs.

Average pressure, $\dfrac{545\cdot55}{12} = 45\cdot46$ lbs.

Thus it will now be quite clear that half a cylinderful of steam, or 6 in. of its length (we are assuming a 12-in. stroke), when expanded to 7 in., or $\frac{7}{6}$ of the volume, becomes $\frac{6}{7}$ of its pressure, and so on.

Referring back to p. 136 we may make use again of the Table of Hyperbolic Logarithms, and the formula there given for finding the mean or average pressure during the whole stroke.

The initial pressure being 55 lbs., and the cut-off being at one-half the stroke, we have (the hyp. log. of ½ being ·693, and R, the ratio of expansion, or ½ inverted, being 2)

$$p, \text{ or the mean pressure} = 55\dfrac{1\cdot693}{2} = 46\cdot55 \text{ lbs.}$$

But this does not agree with the result we just now found, viz., 45·46 lbs. Why? Simply because in the table of pressures we virtually constructed a series of steps, quite correct so far as they went, but a very rude approximation to a curve, such as is expressed by the formula. If the steps, instead of being an inch apart, had been calculated for every half inch, we should be much nearer the truth; but, as students of the calculus very well know, it is only when the steps are taken infinitely close that the error is exhausted, and the serrated line becomes a flowing curve. Hence the difference in the two results.

The left-hand diagram in Fig. 112, traced in a thick black line, is a real diagram taken from a portable engine with a separate cut-off valve and automatic expansion gear. We will follow this black line (which refers only to one end of the cylinder) through a complete revolution. As we have already seen while investigating the action of the slide-valve, the stroke (for each side of the piston separately) is divided into four periods—admission, expansion, exhaust, and compression. These are all distinctly indicated by the diagram before us. The piston is supposed to be moving from left to right, and at the point A the steam pressure behind it reaches to 59 lbs. per square inch, and this is maintained at nearly a constant height until the point B, $4\frac{1}{2}$ in. from the beginning of the stroke, is reached, when the period of admission ceases.

It will be noticed that the point of cut-off is not nearly so sharply defined, on account of the gradual closing of the valve, in the real diagram as in the purely theoretical one to the right hand. From A to B is the line of admission, and from B to C is the line of expansion, approximating to the hyperbolic curve. At C, about $9\frac{3}{4}$ in. from the commencement of the stroke, the exhaust begins to open, and the released steam escapes from the cylinder during the remaining portion of the piston's travel, the rapidly declining pressure being denoted by the short curve C D.

At this point the return stroke begins, and while the piston, urged by a new admission of steam upon the other side, re-traverses the distance D E, the exhaust port remains open, and the pressure in the cylinder, reduced to that of the atmosphere (indicated by the exhaust line D E), coincides with the zero point of the scale. The cylinder, however, still contains steam at atmospheric pressure, and after the piston passes the point E, which marks the closing of the exhaust port, this latent steam undergoes compression for the remainder of the back stroke.

From E to F the pressure of this entrapped steam rises gradually to about 32 lbs., and finally, at F, pre-admission or lead begins. During the last quarter of an inch or so of the stroke, the pressure rises abruptly to 60 lbs., in readiness for the beginning of a new stroke at A. At first sight it would appear that compression, which is a pressure against the advance of the piston, would result in a loss of power. This is not the case, however, as the piston recoils with exactly equal force; and the elastic steam cushion is useful both in absorbing the energy of the reciprocating parts at the changes of direction which occur at the ends of the stroke, and in gradually leading up to the pressure of the steam admitted for the beginning of the next stroke. Indeed, were it not for the compression, the new steam, striking the piston in an empty cylinder some hundreds of times per minute, would speedily hammer the engine to pieces. In locomotive engines, which

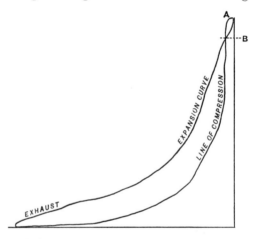

Fig. 113.—Indicator Diagram from a Locomotive Engine, showing the Effect of High Compression.

sometimes attain an enormous piston speed, an extraordinary amount of compression, induced by the early cut-off employed, is found to be in no way injurious, but, on the contrary, highly beneficial.

In some cases this is carried to such an extent that the steam remaining in the cylinder is compressed to above boiler pressure, and a fall takes place when the port is opened. This is illustrated by Fig. 113. Upon the return stroke the pressure rises to the point A, but falls to B directly the port opens, the excess of pressure being discharged into the boiler. This is always indicated by a loop in the corner of the diagram.

THE INDICATOR DIAGRAM

Fig. 114 shows the effect of *insufficient passage area*, or port opening. This particular defect is known as wire drawing, and is caused by the inability of the steam to make its way into the

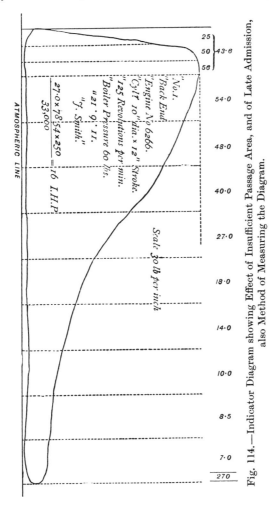

Fig. 114.—Indicator Diagram showing Effect of Insufficient Passage Area, and of Late Admission, also Method of Measuring the Diagram.

cylinder fast enough to maintain the initial pressure behind the rapidly advancing piston. The pressure, it will be observed, has fallen to such a degree that it is hardly possible to distinguish where the admission line ends and the expansion curve begins.

Another defect is illustrated in the same diagram, namely, *late admission*. There is no compression, and the admission of steam is so gradual that the piston has travelled some distance before the full pressure is arrived at.

Back pressure is also shown by the exhaust line being a pound or two above the atmospheric line. The late admission, in the engine from which this diagram was taken, could easily have been cured by moving the eccentric further round upon the shaft in the direction in which the engine was running, thus giving more *angular advance*; the exhaust would then be opened and closed earlier than at present. The back pressure would be reduced and the compression increased by this means, but the falling admission line is inseparable from the method of governing by throttling the steam inlet, though this is a particularly bad case of it.

Compare the diagram (left hand) in Fig. 112, where the well-

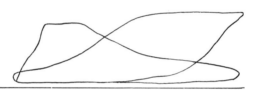

Fig. 115.—Pair of Diagrams, showing Unequal Valve-setting.

sustained admission line shows that the governing has been effected by automatic expansion gear, the boiler pressure being in unrestricted communication with the piston, up to the point at which the expansion valve cuts off the steam altogether.

As before mentioned, the diagram only shows the pressure upon one side of the piston. It is highly necessary that both sides should be indicated. Fig. 115, for instance, shows the effect of *unequal valve-setting*; the valve spindle or rod is either too long or too short, and its adjustment would equalise the diagrams. The right-hand diagram shows an undue amount of compression and of lead, while the diagram to the left is deficient in both respects as well as in the cut-off, and both suffer from an injurious amount of back pressure.

We mentioned a moment ago that the indicator diagram only shows the pressure upon one side of the piston, but this is true only in a conventional sense; the compression line, showing pressure *against* the piston, should, strictly speaking, take its place at the toe-end of the diagram, as in Fig. 87, p. 113. For all ordinary purposes, if the compression lines at both ends of the cylinder are of equal value, they may be exchanged without sensible error, but it is just as well to remember that this is only a convention and not the reality. The critical examination of the various lines of the indicator diagram is a most interesting and useful study, and the reader who desires further guidance is referred to Mr Pickworth's book, "The Indicator Diagram," published by Emmott & Co., Ltd., Manchester.

Our remaining observations must be devoted to the means of deducing the *power* of the engine from the indicator diagram, and to a concise description of the instrument itself; the calculations of the actual quantity of steam, and of heat, expended in producing the power do not lie within the scope of the present volume, and must reluctantly be passed over. The term *nominal horse-power* is utterly misleading, and is simply a commercial expression by which portable engines are bought and sold. *Brake* or *effective* horse-power is the criterion of value for money; the indicator diagram gives the power developed *in the cylinder* fairly accurately (if the instrument is used under proper conditions), but from this result has to be subtracted the power consumed in driving the engine itself, which in some cases is unexpectedly great. To make this quite clear, suppose an observation of the indicated horse-power shows 50 I.H.P.; the brake horse-power, or the power "available for distribution," would be less than this by the *engine friction*, which might be anything from 10 to 20 per cent. Hence the *efficiency* of an engine is expressed as a fraction showing the percentage, thus :—

$$\frac{\text{B.H.P.} \times 100}{\text{I.H.P.}} = \text{efficiency per cent.}$$

But, to return to our diagram, the indicator consists in principle of a small cylinder having within it an accurately-fitting piston (usually of one-half a square inch area, or ·7979 in. diameter), the under side of which is in free communication with one end of the engine cylinder. The piston rod of the indicator at its upper end carries a multiplying lever with a pencil arranged so as to mark upon a card, which slides backwards and forwards in front of it in unison with the movement of the engine piston. A stiff spiral spring, adapted to the pressure likely to be met with in the engine cylinder, is placed above the little indicator piston, so that the height of the latter, being dependent upon the compression of the spring, is an exact measure of the pressure acting beneath it.

If now the cock between the indicator and the engine cylinder be closed, the pencil, remaining stationary at its lowest point, will trace upon the moving card a *horizontal* line, and if the atmosphere be admitted under the indicator piston, this line will denote atmospheric pressure. If the cock be opened to the cylinder, and the card stopped, a *vertical* line will be drawn, the upper and lower extremities of which record the highest and lowest pressures respectively which have occurred during the stroke of the engine. The combined result of the simultaneous movements of card *and* pencil is the delineation upon the former of a diagram more or less approximating to those we have been considering. The pencil is only allowed to touch the card while one revolution is being performed.

Fig. 116 shows several stages in the tracing of the diagram during one stroke, or half a revolution.

In No. 1 the card is just beginning to move in the direction of the arrow, and the piston rod of the indicator has shot right up to the top, showing that the full pressure of the steam has just been admitted to the cylinder. No. 2 shows the admission line completed, and the remainder of the figures show various stages of the expansion and exhaust lines. In No. 7 the card is just ready to move with the return stroke in the direction of the arrow, so as to draw the return exhaust line. The atmospheric line is visible in all the views, having been drawn before the diagram was begun.

The indicator nowadays takes many forms, and it is matter for regret that the exigencies of space prevent a discussion of the various types now in use.

Modern instruments almost without exception, though varying in detail, embody the principle of a very short and stiff piston spring with a very light multiplying lever and parallel motion for carrying the pencil. The card (really a slip of prepared metallic paper which can be marked by a brass point) is wrapped round a drum, which, by means of a cord and some form of reducing motion,

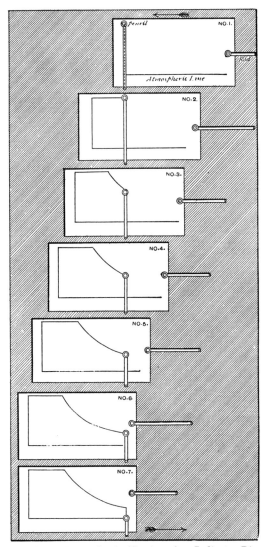

Fig. 116.—Seven Stages in the Tracing of an Indicator Diagram.

is partially rotated in unison with the stroke of the engine piston against the pull of a coiled spring inside the drum.

Indicated Horse-Power.—In a steam engine the power exerted is always equal to the resistance overcome, and we can always measure that by ascertaining the mean pressure upon the piston and its speed. Total pounds of pressure upon the piston multiplied by the number of feet it is driven through in one minute would evidently be exactly balanced by the same number of pounds'

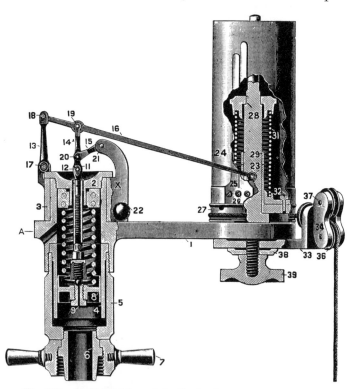

Fig. 117.—Sectional View of the Crosby Steam-Engine Indicator.

weight lifted through an equal number of feet in the same time, supposing the engine to work without friction, and to be arranged for winding a weight up, as in a colliery engine. The actual work of the engine may be to draw a train, propel a steamship, or drive a single machine or a whole factory, but the work done can always be reduced to the unit of *foot-pounds* (or pounds raised one foot high) *in one minute.* The unit of 1 ft.-lb. being rather in-

THE INDICATOR DIAGRAM

conveniently small, a multiple of this called the horse-power, of 33,000 ft.-lbs., is adopted.

The expression for horse-power, therefore, is :—

$$\text{H.P.} = \frac{\text{lbs. per sq. in.} \times \text{area of piston} \times \text{feet per minute}}{33,000},$$

and the indicator diagram, as a means of measuring the power of the engine, may be considered simply as a method of arriving at the actual average or mean pressure upon the piston during the whole stroke. The *height* of the space included within the lines of the diagram, as before explained, represents at any given point the pressure within the engine cylinder at that point. The mean height of the diagram, therefore (as measured by the scale provided corresponding with the spring in use), will give the mean cylinder pressure, which is what we require for the first factor in the expression above. The mean height of an irregular plane figure, as an engine diagram, may be ascertained, more or less approximately, by dividing it up into a number of compartments, and measuring the height of each one separately; and this is the method commonly used (in the absence of a special instrument, called the planimeter) with indicator diagrams. The closer these divisions are taken, the nearer the approach to accuracy in the result, but ten are usually taken. The diagram, Fig. 114, is so divided, but because of the slope of the left-hand line of the figure, the first compartment containing it has been further subdivided into three spaces, and the mean of three measurements, 43·6 lbs., taken as the mean height. The second and succeeding compartments are each measured once only, the results being carried out to the margin so as to stand vertically over one another, for ease in adding up. The total, 270·1, by removing the decimal point once to the left, becomes the mean 27·0 (the ·01 may be neglected). Then as a 12-in. stroke and 125 revolutions per minute give 250 ft. of piston, and assuming a 10-in. diameter cylinder (area 78·54 sq. in.), we have :—

$$\text{I.H.P.} = \frac{27 \times 78\cdot54 \times 250}{33,000} = 16\cdot06,$$

or, say, 16 I.H.P. In this we are assuming that the diagram represents correctly both ends of the cylinder, but a reference to

Fig. 115 will show that this should not be taken for granted. Both ends should be indicated separately, and the mean of the two diagrams taken as the real mean pressure. (A three-way cock under the indicator, with pipes extending to both ends of the cylinder, enables both diagrams to be taken upon the same card, and their similarity or otherwise compared.) The scale corresponding with the spring used should be marked upon the diagrams, and the further information shown in Fig. 114 should also be written upon the card for future reference.

When a large number of diagrams have to be taken from the same engine, much time in measurement and calculation may be saved by adopting two expedients. First, instead of measuring each compartment separately with the scale, a long strip of paper may be applied to each point, and the heights ticked with a pencil continuously along it. The number of inches of total length can then be measured with a 2-ft. rule, the scale being only employed for the last fraction of an inch. Secondly, a good deal of the calculation may be done once for all by preparing a constant thus:—

$$C = \frac{\text{area of piston} \times \text{stroke in inches}}{33{,}000 \times 6},$$

which will be a decimal fraction representing the horse-power of that particular engine at 1 lb. pressure and 1 ft. of piston speed per minute, only requiring to be multiplied by the average pressure taken from each diagram, and by the revolutions per minute observed simultaneously, to give at once the indicated horse-power. If the revolutions per minute be so nearly uniform that they may be regarded as invariable, this factor may also be included in the constant, which will then only require to be multiplied by the mean pressure of each diagram to give the result. In our case the constant would be—

$$C^1 = \frac{78 \cdot 54 \times 250}{33{,}000} = \cdot 595,$$

which only requires to be multiplied by the 27 lbs. of the diagram to give the horse-power, 16·06. It is a waste of time, as a rule, to give more than one place of decimals to the number representing indicated horse-power.

THE INDICATOR DIAGRAM 163

The Friction Brake, by which the effective or brake horse-power (B.H.P.) of an engine is ascertained, in its elementary form is a very simple apparatus, and may be described in equally simple language, thus:—

Suppose the flywheel of a portable engine to be encircled by a band, lined with wood blocks, and capable of being screwed up so as to grip the wheel to any desired extent. If now this is anchored so that it cannot turn with the wheel, it is obvious that we can put any load we please on the engine up to the point of stopping it altogether. If we substitute a weight, say, of W lbs. for the fixed anchorage, and so adjust the screw that this weight remains suspended, neither resting on the ground nor being carried round by the flywheel, we can calculate the load on the engine by the same method as if it were winding that load out of a pit at a rate corresponding with the revolutions of the engine multiplied by the circumference of an imaginary drum whose radius is just equal to the horizontal distance between the centre of gravity of the suspended weight and the centre of the flywheel.

Call this distance l, and let it be $2\frac{1}{2}$ ft. in this case, and let R, the number of revolutions per minute, be 125, then we have, if we call the weight W 16·8 lbs., the following, as the expression for B.H.P. or brake horse-power:—

$$\text{B.H.P.} = \frac{2 \times 2\frac{1}{2} \times 3\cdot1416 \times 125 \times 16\cdot8}{33,000} = 1,$$

or, generally, $$\text{B.H.P.} = \frac{2l \times \pi \times R \times W}{33,000}.$$

That is to say, that it requires exactly 1 B.H.P. to lift a load of 16·8 lbs. through 1963·5 ft. per minute. This is clear enough, but (the reader will say) the mere suspension of a weight—holding it at arm's length, so to speak—is a very different thing from lifting it against the pull of gravity through nearly 2,000 ft. per minute. Exactly: and yet the same formula is correct for both cases—that is to say, if you had two engines, one lifting the load by a winding drum, and the other holding the same load suspended at an equal radius, the indicated horse-power developed would be precisely the same in both cases. The explanation is not far to seek, and it is this: you might wind for any length of time without increasing the sensible temperature of the drum or of the rope, but a ten minutes' run on the brake would probably so heat up the rim of the wheel that by its expansion it would part company with the spokes and a general smash would ensue.

In practice, a wheel is used having a rim of channel section. Water is fed into the rim at one side; and a short piece of pipe, or a scoop, fixed so as to dip into the revolving mass, peels the water out on the other side. Put the hand under this scoop, and you will be surprised to find that the issuing water is scalding hot.

If, now, all the heat so generated, together with that dissipated into the air as the wheel revolves, could be collected and measured, it would be found that a definite relation exists between these two forms of energy—heat and mechanical power—which may be stated thus:—

Heat and mechanical energy are mutually convertible; and heat requires for its production, or produces by its disappearance, mechanical energy in the proportion of 772 ft.-lbs. for 1 unit of heat.

Thus each of these two forms of energy may be expressed in terms of the other: a quantity of heat may be reduced to, and expressed in, foot-pounds, by multiplying it by 772; and a quantity of work may be equally well expressed in foot-pounds or in heat-units. For instance, a horse-power (or 33,000 ft.-lbs. of work per minute) may be stated in heat measure by saying that 1 H.P. is equivalent to $\dfrac{33,000}{772} = 42\cdot 7$ heat-units per minute.

The determination of the actual mechanical equivalent of heat is one of the monuments of the nineteenth century; and the letter J, the mathematical symbol of the number 772, is an enduring memorial to the man who devoted his life's work to its evaluation.

The exigencies of space (and of time) prevent further discussion of the interesting subject of the friction-brake here; but those who desire more practical treatment of the question are referred to a recent article by the author in *Cassier's Magazine* for September 1911, where various types of engine-testing brakes are described and illustrated.

THE INDICATOR DIAGRAM

Our task is now completed. What the future may have in store for the portable engine remains to be seen. Whether or not steam engines will be superseded by the internal combustion engine as a motive force, the motor on wheels will, it can hardly be doubted, always remain with us in some form or other until the demand for power shall cease for ever.

THE END.

INDEX

NAMES

BACH, 14.
  Barrett, 17.
Biddell & Balk, 107.
Boulton & Watt, 4.
Brown & May, 39-41, 68, 69.
Burrell, 17.
Butlin, 17.

CABRON, 17.
  Cambridge, 8, 9, 11.
Cartwright, 2.
Clayton & Shuttleworth, 8, 10, 13, 17, 18, 20-22, 41-47, 68-71.
Crosby, 160.

DAVEY, PAXMAN, & CO., 49, 50, 71-75.
Davies, 7.
Dean, 8, 9, 12.
Dewrance, 110.

FOSTER, WM., & CO., 50, 75-78.

GARRETT, R., & SONS, 17, 18, 33, 51-55, 78-81, 99-101.

HARTNELL, 64, 81, 82.
  Hawkins, 4.
Hensman, 17.
Hornsby, 12, 16, 17, 22, 55.
Hutton, 66.

LEOPOLD of Plunitz, 1.

MARSHALL, SONS, & CO., 36, 55, 56, 81, 82, 102.

Murdoch, 2.
Murray, 2.

OGG, 12.

PICKARD, 2.
  Pickering, 41, 46, 47, 49, 51, 53, 54, 56, 57, 71, 77, 78, 82, 85, 90, 102.
Pickworth, 157.

RANSOMES, SIMS, & JEFFERIES, 7, 18, 57, 82-84, 107, 108.
Reading Iron Works, 21.
Richardson, 21, 35.
Rider, 78, 80, 90, 92, 102.
Robey & Co., 19-22, 35, 57-60, 85-90, 97, 103, 104, 106.
Roe, 17.
Royal Agricultural Society, 4, 6-21, 50, 62, 65, 75, 79.
Ruston, Proctor, & Co., 60-62, 90-95, 98, 102, 105.

SMEATON, 1.

TREVITHICK, 3.
  Turner, E. R. & F., 62-64.
Tuxford, 6, 14-17, 21, 22.

WASBROUGH, 2.
  Watt, 2.
Webb, 20.

ZEUNER, 137-149.

INDEX

SUBJECT-MATTER

ADJUSTING governor, 122-124.
Angular advance, 131, 132, 133, 134, 156.
Angular bearings, 51, 53, 79.
Annealing boiler plates, 31.
Automatic expansion gear, 24, 26, 71, 73, 74, 77, 78, 80, 81, 84, 85, 88-90, 92, 93, 102, 117, 118, 120-122, 153, 156.

BACK pressure, 135, 154, 156.
Blast-pipe, 3.
"Blow through," 116.
Boiler plates, 30.
Boiler stay-bolts, 33, 37, 38.
Brake, friction, 163-165.
Brake horse-power, 66, 67, 157, 163-165.
"Britannia" boiler, 102.

CARE of boilers, 128, 129.
Cast-iron saddle, 28, 41, 79.
Circular firebox, 102-107.
Coal analysis, 125, 126.
Compound portable engines, 26, 65-95, 103.
Compression, 112, 113, 134, 139, 140, 147, 149, 153, 154, 156, 157.
Condensing engines, 1, 97-99.
Connecting rod, angularity of, 119, 139, 142, 143.
Crank diagram, 131-133.
Crankpin, 112, 114, 115.
Crankpin, pressures on, 112, 113.
Crankshaft governor, 21, 64.

DEAD centres, 112-116, 120, 121, 133, 137, 142, 144.

Detachable engine, 28, 29, 49, 71, 82.
Diagram, crank, 131-133.
Diagram, indicator, 24-26, 73, 77, 90, 94, 113, 150-162.
Diagram, Zeuner, 137-149.

ECONOMY of compounding, 66, 67.
Effective horse-power (see Brake Horse-power).
Equalising diagrams, 119, 120, 156.
Exhaust lap, 139, 140, 141, 147.
Expansion, table of, 136.
Expansion valve, 12, 25.

FEED pump, 18, 54, 55.
Feed-water heater, 3, 12, 54, 55, 78.
Firebox crown plate, 31-37.
Firebox roof stays, 31-37.
Firing, 124-128.
Flanging boiler plates, 22, 30.
Flywheel energy, 123.
Flush-topped boiler, 37, 38, 41, 78.
Forced oil feed, 56, 101, 102, 111.
Formula for lap and lead, 135.
Friction brake, 163-165.

GOVERNOR expansion gear, 21, 47, 49, 53, 54, 56, 61, 64, 81, 82, 122-124.
Governor, throttle-valve, 25, 26, 41, 46, 47, 49, 51, 53, 54, 56, 57, 71, 77, 78, 82, 85-90, 102.
Guide bars, 111, 112, 114.

INDEX

HIGH - PRESSURE portable engine, 101, 102.
Hot bearing, 109-111.
Hyperbolic logarithms, 136.

INDICATED horse-power, 157, 160-162.
Indicator, 157-162.
Indicator diagrams, 24-26, 73, 77, 90, 94, 113, 120, 150-162.

"**K**NOCKING," 114, 115.

LAGGING, 41, 129.
Lap, 132-136.
Lead, 133-136.
Leaking pistons, 116.
Leaking valves, 116.
Logarithms, hyperbolic, 136.
Lubrication, 56, 109-111.

NOMINAL horse-power, 39, 157.

OIL pump, mechanical, 101, 102, 111.
Outside steam chests, 71, 82.
Outside steam inlet, 23, 41, 55, 57, 60, 68, 82.

PISTON, 115, 116.

REMOVABLE firebox, 103, 107.
Riveted joints, 22, 30, 31.
Rivet steel, 30.

SECTIONAL view of cylinder, 44, 61, 88, 89, 92, 119, 130, 142.
Semi-portable engines, 26, 96, 101.
Setting engine square, 114.
Setting expansion valves, 117, 118.
Setting slide-valves, 117-122.
Slide-valve, 2, 12, 116-122.
Slide-valve proportions, 130-134, 140-150.
Slide-valve sections, 130, 134, 140, 144-147.
Sliding plummer-blocks, 23, 24, 29, 41, 45, 49, 56, 57, 64, 81, 82, 102.
Stay-rods, 23, 24, 29, 37, 38.
Steam-heated stay, 61, 62, 93, 102.
Steam jacket, 18, 23.
Steel boiler plates, 30.
Steel rivets, 30.
Sun-and-planet motion, 2.
Superheater, 100, 101.

TABLES of dimensions, 17, 68, 75, 77, 84, 85, 93, 134.
Table of pressures, 152.
Tail rods, 79.
Testing valves and pistons, 116.
Thermo-dynamic law, 164.
Throttle-valve, 25.
Tie-rods (*see* Stay-rods).
Trunk guides, 13, 46, 93.

VARIABLE eccentric 45, 46, 62, 148, 149.
Variable expansion gear (*see* Governor expansion gear).

ZEUNER diagram, 137-149.